Synthesis Lectures on Engineering, Science, and Technology

The focus of this series is general topics, and applications about, and for, engineers and scientists on a wide array of applications, methods and advances. Most titles cover subjects such as professional development, education, and study skills, as well as basic introductory undergraduate material and other topics appropriate for a broader and less technical audience.

Ulziibayar Vandandoo · Tugal Zhanlav ·
Ochbadrakh Chuluunbaatar ·
Alexander Gusev · Sergue Vinitsky ·
Galmandakh Chuluunbaatar

High-Order Finite Difference and Finite Element Methods for Solving Some Partial Differential Equations

 Springer

Ulziibayar Vandandoo
School of Applied Sciences, Mongolian
University of Science and Technology
Ulaanbaatar, Mongolia

Ochbadrakh Chuluunbaatar
Meshcheryakov Laboratory of Information
Technologies
Joint Institute for Nuclear Research
Dubna, Russia

Institute of Mathematics and Digital
Technology
Mongolian Academy of Sciences
Ulaanbaatar, Mongolia

Sergue Vinitsky
Bogoliubov Laboratory of Theoretical Physics
Joint Institute for Nuclear Research
Dubna, Russia

Peoples' Friendship University of Russia
(RUDN University)
Moscow, Russia

Tugal Zhanlav
Institute of Mathematics and Digital
Technology
Mongolian Academy of Sciences
Ulaanbaatar, Mongolia

Alexander Gusev
Meshcheryakov Laboratory of Information
Technologies
Joint Institute for Nuclear Research
Dubna, Russia

Dubna State University
Dubna, Russia

Galmandakh Chuluunbaatar
Meshcheryakov Laboratory of Information
Technologies
Joint Institute of Nuclear Research
Dubna, Russia

Peoples' Friendship University of Russia
(RUDN University)
Moscow, Russia

ISSN 2690-0300   ISSN 2690-0327 (electronic)
Synthesis Lectures on Engineering, Science, and Technology
ISBN 978-3-031-44786-0   ISBN 978-3-031-44784-6 (eBook)
https://doi.org/10.1007/978-3-031-44784-6

This Springer imprint is published by the registered company Springer Nature Switzerland AG
The registered company address is: Gewerbestrasse 11, 6330 Cham, Switzerland

Paper in this product is recyclable.

Preface

The monograph is devoted to the construction of the high-order finite difference and finite element methods for numerical solving multidimensional boundary-value problems (BVPs) for different partial differential equations, in particular, linear Helmholtz and wave equations, nonlinear Burgers' equations, and elliptic (Schrödinger) equation. Despite of a long history especially in development of the theoretical background of these methods there are open questions in their constructive implementation in numerical solving the multidimensional BVPs having additional requirement on physical parameters or desirable properties of its approximate solutions.

Over the last two decades many papers on this topics have been published, in which new constructive approaches to numerically solving the multidimensional BVPs were proposed, and its highly desirable to systematically collect these results. This motivate us to write thus monograph based on our research results obtained in collaboration with the co-authors. Since the topic is importance we believe that this book will be useful to readers, graduate students and researchers interested in the field of computational physics, applied mathematics, numerical analysis and applied sciences.

The monograph contains three chapters and two appendixes.

In Chap. 1, we present accurate finite difference methods (FDMs) for the two-dimensional Helmholtz and one-dimensional wave equations, using Newton's method or its continuous analogue described in Appendix A. A new exact FDM for solving the two-dimensional Helmholtz equation is also presented. The main features of this method are that it can be applied to solve the two-dimensional Helmholtz equation for any wavenumber without using a fine mesh.

In Chap. 2, we construct new stable and high-order FDMs to solve one- and two-dimensional Burgers' equations, as well as two-dimensional coupled Burgers' equations with a corresponding initial condition and boundary conditions. The proposed high-order FDMs are applied to the calculation of several exactly solvable examples. The numerical results are in good agreement with the exact solutions in a wide range of the Reynolds number values and confirm the approximation orders of the proposed methods.

In Chap. 3, we construct newstable and high-order finite element methods (FEMs) to solve multidimensional BVPs for the elliptic (Schrödinger) and Helmholtz equations that preserved continuous derivative of the FEM solution on boundaries of the joint finite elements. The efficiency of the applied high-order FEMs is shown by benchmark calculations of several exact solvable examples for Helmholtz problems, and the BVP with table value coefficients in a two-dimensional domain describing the quadrupole-octupole vibrational collective nuclear model. The PI-type fully symmetric Gaussian quadrature rules for evaluation of FEM multidimensional integrals are constructed in Appendix B.

Ulaanbaatar, Mongolia	Ulziibayar Vandandoo
Ulaanbaatar, Mongolia	Tugal Zhanlav
Dubna, Russia	Ochbadrakh Chuluunbaatar
Dubna, Russia	Alexander Gusev
Dubna, Russia	Sergue Vinitsky
Dubna, Russia	Galmandakh Chuluunbaatar

Acknowledgements We particularly thank Professor V.L. Derbov, Professor A. Góźdź and Professor A. Dobrowolski for fruitful collaboration. The monograph was supported partially by the Foundation of Technology of Mongolian University of Science and Technology theme research project MFUND-03/2022 "A study of the symbolic and numerical methods for solving of nonlinear system", the Hulubei–Meshcheryakov Joint Institute for Nuclear Research program, the Russian Foundation for Basic Research and the Ministry of Education, Culture, Science and Sports of Mongolia (No. 20–51–44001), the Peoples' Friendship University of Russia (RUDN) Strategic Academic Leadership Program (No. 021934-0-000) and the grant from the Ministry of Education and Science of Mongolia (No. ShuG 2021/137).

Contents

List of Figures

List of Tables

Accurate Finite-Difference Methods for Helmholtz and Wave Equations

1

Abstract

Accurate finite difference methods (FDMs) for the numerical solution of two-dimensional Helmholtz and one-dimensional wave equations are proposed. The accurate finite difference equations and the boundary conditions are formulated as algebraic and algebraic-eigenvalue problems. Calculation schemes for solving these problems using conventional numerical methods and Newton's method or its continuous analogue, and the corresponding test exactly solvable examples are presented and analyzed. A new exact FDM for solving the two-dimensional Helmholtz equation is also presented. This method is implemented by solving two one-dimensional problems under the separation of variables restriction. The main feature of this method is that it can be applied for solving the two-dimensional Helmholtz equation with any wavenumber without using a fine mesh. The method accuracy is analyzed theoretically. The effectiveness and performance of the proposed FDMs are demonstrated by comparing known methods with several examples.

1.1 Introduction

The FDM is a standard numerical method for solving BVPs. Recently, considerable attention has been paid to the construction of an accurate (or exact) finite difference approximation for some ordinary and partial differential equations [1–3]. Accurate FDMs are developed for Helmholtz and wave equations with general boundary conditions (including the initial conditions for the wave equation) on the rectangular domain in $\mathbb{R}^2$ [4, 5]. The methods proposed here come out from [6] and is based on separation of variables for a sufficiently smooth solution. The efficiency and accuracy of the method are tested on several examples.

© The Author(s), under exclusive license to Springer Nature Switzerland AG 2024
U. Vandandoo et al., *High-Order Finite Difference and Finite Element Methods for Solving Some Partial Differential Equations*, Synthesis Lectures on Engineering, Science, and Technology, https://doi.org/10.1007/978-3-031-44784-6_1

The Helmholtz equation arises from time-harmonic wave propagation, and its solutions are often required in many applications such as aero- and hydroacoustics, electromagnetic wave scattering, and geophysical problems. Different FDMs, including standard central finite difference ones and new methods with higher accuracy, in particular, the exact one, are used to solve the Helmholtz equation. It should be noted that all the numerical methods mentioned above require a very fine mesh to ensure the accuracy of the computed solutions at high wavenumbers k. The reason for the pollution effect is that the solution of the Helmholtz equation at a high wavenumber k is highly oscillatory. To accurately capture the oscillatory behavior, it is necessary that kh be small, where h is the mesh size of discretization.

It is known [7] that, for $kh = \text{const}$, the errors of the numerical solutions grow rapidly as the wavenumber k increases. Hence, for high wavenumber problems, the pollution effect can only be avoided when using a very fine mesh. In the one-dimensional case, it is shown that the pollution effect can be completely eliminated. However, it is shown (see [7] and references therein) that the pollution effect is inevitable when applying the FDM to multidimensional systems.

In [8], exact FDMs are proposed for solving the Helmholtz equation at any wavenumber, and it is pointed out that the most important feature of the new FDMs is that, although the resulting linear system has the same simple structure as those derived from the standard central difference method, the technique is capable of solving the Helmholtz equation at any wavenumber without using a fine mesh. However, for two-dimensional problems, these new FDMs are not effective unless information about the angle is known. Recently, in [7] the pollution effect of the discrete singular convolution algorithm has been investigated, and it has been demonstrated that this algorithm can be a pollution-free method for solving the Helmholtz equation. It is clear from the above that solving the Helmholtz equation at high wavenumbers numerically remains one of the most difficult tasks in scientific computing.

1.2   Accurate Finite Difference Methods for the Helmholtz Equation

1.2.1   Statement of the Problem

Let $\Omega = (a, b) \times (c, b)$ be an open rectangular domain in Euclidean $\mathbb{R}^2$ space with boundary given by $\partial\Omega$. The aim is to determine a function $u(x, y)$, satisfying equation [4]

$$\left(\frac{\partial^2}{\partial x^2} + \frac{\partial^2}{\partial y^2} \right) u(x, y) + Cu(x, y) = 0, \quad (x, y) \in \Omega, \tag{1.1}$$

with boundary conditions

$$\alpha_1 u(a, y) - \beta_1 \left. \frac{\partial u(x, y)}{\partial x} \right|_{x=a} = \xi_1(y),$$

$$\alpha_2 u(b, y) + \beta_2 \left. \frac{\partial u(x, y)}{\partial x} \right|_{x-b} = \xi_2(y),$$

$$\alpha_3 u(x, c) - \beta_3 \left. \frac{\partial u(x, y)}{\partial y} \right|_{y=c} = \xi_3(x), \tag{1.2}$$

$$\alpha_4 u(x, d) + \beta_4 \left. \frac{\partial u(x, y)}{\partial y} \right|_{y=d} = \xi_4(x),$$

where C, α_i and β_i, $i = 1, \ldots, 4$, are given numbers, and $\xi_1(x)$, $\xi_2(x)$, $\xi_3(y)$ and $\xi_4(y)$ are given smooth functions.

It is well known that the stabilized oscillation problems and diffusing processes in gas lead to the so called Helmholtz equation (1.1) with a positive coefficient $C = \lambda^2$. The diffusing process in the moving field leads to the Eq. (1.1) with negative coefficient $C = -\lambda^2$. If $C = 0$ the Eq. (1.1) leads to Laplace's ones. Obviously, the properties of the solution of Eq. (1.1) depend essentially upon the sign of the problem (1.1), (1.2) has an unique and sufficiently smooth solution.

By virtue of separation of variables method looking for the solution $u(x, y)$ of Eqs. (1.1), (1.2) in the form

$$u(x, y) = U_1(x)U_2(y), \tag{1.3}$$

we arrive to equation

$$\frac{U_1''(x)}{U_1(x)} + \frac{U_2''(y)}{U_2(y)} = -C, \tag{1.4}$$

that splits into two independent equations

$$U_1''(x) = \omega U_1(x), \tag{1.5}$$

and

$$U_2''(y) = \gamma U_2(y), \quad \gamma = -C - \omega, \tag{1.6}$$

where the unknown separation constant ω is to be found. By virtue of (1.3) the boundary condition (1.2) is split for $U_1(x)$ and $U_2(y)$

$$\alpha_1 U_1(a) - \beta_1 U_1'(a) = \chi_{10},$$
$$\alpha_2 U_1(b) + \beta_2 U_1'(b) = \chi_{2N}, \tag{1.7}$$

and

$$\alpha_3 U_2(c) - \beta_3 U_2'(c) = \chi_{30},$$
$$\alpha_4 U_2(d) + \beta_4 U_2'(d) = \chi_{4N}. \tag{1.8}$$

When $\omega \geq 0$, the solution of boundary-value problem (BVP) (1.5), (1.7) is found in a closed form

$$U_1(x) = \frac{\chi_{10} F_1(\omega) + \chi_{2N} F_2(\omega)}{F_3(\omega) + \sqrt{\omega} F_4(\omega)}, \tag{1.9}$$

where

$$F_1(\omega) = \alpha_2 \sinh(\sqrt{\omega}(b - x)) + \beta_2 \sqrt{\omega} \cosh(\sqrt{\omega}(b - x)),$$
$$F_2(\omega) = \alpha_1 \sinh(\sqrt{\omega}(x - a)) + \beta_1 \sqrt{\omega} \cosh(\sqrt{\omega}(x - a)),$$
$$F_3(\omega) = (\alpha_1 \alpha_2 + \omega \beta_1 \beta_2) \sinh(\sqrt{\omega}(b - a)), \tag{1.10}$$
$$F_4(\omega) = (\alpha_1 \beta_2 + \alpha_2 \beta_1) \cosh(\sqrt{\omega}(b - a)).$$

When $\omega < 0$ the functions sinh and cosh in (1.9), (1.10) are to be replaced by sin and cos, respectively, and $\sqrt{\omega}$ replaced by $\sqrt{-\omega}$. Analogously, we can find the solutions of BVP (1.6) and (1.8) in closed form. Then from (1.3) and (1.9) clear, that the problem consists in determining the separation constant ω.

1.2.2 Construction of the Accurate Finite Difference Equations

For the numerical solution of problem (1.1), (1.2) is introduced the uniform rectangular grid Ω_h:

$$\Omega_h = \{(x_i, y_j) | x_i = x_0 + i h_1, \ y_j = y_0 + j h_2, \ i = 0, 1, \ldots, N, \ j = 0, 1, \ldots, M\}, (1.11)$$

where $h_1 = (b - a)/N$ and $h_2 = (d - c)/M$ are the mesh sizes in the x and y directions, respectively. Usually, the Eq. (1.1) is approximated by the five-point difference equation

$$\frac{y_{i+1,j} - 2y_{i,j} + y_{i-1,j}}{h_1^2} + \frac{y_{i,j+1} - 2y_{i,j} + y_{i,j-1}}{h_2^2} + C y_{i,j} = 0, \tag{1.12}$$
$$i = 1, \ldots, N - 1, \quad j = 1, \ldots, M - 1.$$

The local discretization error of the Eq. (1.12) is of $O(h_1^2 + h_2^2)$ order. Now we describe how to derive the accurate difference method for Eq. (1.1). To this end, we consider expression

$$(\Lambda_1 + \Lambda_2) u_{i,j} = \frac{u_{i+1,j} - 2u_{i,j} + u_{i-1,j}}{h_1^2} + \frac{u_{i,j+1} - 2u_{i,j} + u_{i,j-1}}{h_2^2}, \tag{1.13}$$

where $u_{i,j} = u(x_i, y_j)$. If we donate by U_{1i} and U_{2j} the values of $U_1(x_i)$ and $U_2(y_j)$, respectively, the using (1.3) the Eq. (1.13) maybe written as

$$(\Lambda_1 + \Lambda_2)u_{ij} = U_{2j}\Lambda_1 U_{1i} + U_{1i}\Lambda_2 U_{2j}. \tag{1.14}$$

Due to smoothness assumption of solution $u(x, y)$, as well as, functions $U_1(x)$ and $U_2(y)$, the Taylor series expansion yields

$$\Lambda_1 U_{1i} = U_1''(x_i) + 2\sum_{k=1}^{\infty} \frac{h_1^{2k} U_1^{(2k+2)}(x_i)}{(2k+2)!}, \tag{1.15}$$

$$\Lambda_2 U_{2j} = U_2''(y_j) + 2\sum_{k=1}^{\infty} \frac{h_2^{2k} U_2^{(2k+2)}(y_j)}{(2k+2)!}. \tag{1.16}$$

Because of (1.5) we have

$$U_1^{(2k)} = \omega^k U_1, \quad U_2^{(2k)} = \gamma^k U_2, \quad k = 1, 2, \ldots. \tag{1.17}$$

Taking into account (1.15)–(1.17) in (1.14) it follows that

$$\left(\Lambda_1 + \Lambda_2 + C - 2\sum_{k=1}^{\infty} \frac{h_1^{2k}\omega^{k+1} + h_2^{2k}\gamma^{k+1}}{(2k+2)!}\right)u_{ij} = 0, \tag{1.18}$$

$$i = 1, 2, \ldots, N-1, \quad j = 1, 2, \ldots, M-1.$$

The difference equation (1.18) contains unknown nonzero parameter ω and therefore it may be considered as a nonlinear equation with respect to the parameter ω and u_{ij}. The series in (1.18) may be expressed through analytical functions depending on the sign of quantities ω and β and thereby the Eq. (1.18) can be written as

$$(\Lambda_1 + \Lambda_2 - 2D(\omega))u = 0, \quad (x, y) \in \Omega_h. \tag{1.19}$$

There are three cases

1. Let $C = \lambda^2 > 0$. Then

$$D(\omega) = \begin{cases} \dfrac{\cos(\sqrt{-\omega}h_1) - 1}{h_1^2} + \dfrac{\cosh(\sqrt{\gamma}h_2) - 1}{h_2^2}, & \omega \in (-\infty, -\lambda^2), \\[2ex] \dfrac{\cos(\sqrt{-\omega}h_1) - 1}{h_1^2} + \dfrac{\cos(\sqrt{-\gamma}h_2) - 1}{h_2^2}, & \omega \in [-\lambda^2, 0), \\[2ex] \dfrac{\cosh(\sqrt{\omega}h_1) - 1}{h_1^2} + \dfrac{\cos(\sqrt{-\gamma}h_2) - 1}{h_2^2}, & \omega \in [0, +\infty). \end{cases} \tag{1.20}$$

2. Let $C = 0$. In this case D is given by

$$D(\omega) = \begin{cases} \dfrac{\cos(\sqrt{-\omega}h_1) - 1}{h_1^2} + \dfrac{\cosh(\sqrt{-\omega}h_2) - 1}{h_2^2}, & \omega < 0. \\[2ex] \dfrac{\cosh(\sqrt{\omega}h_1) - 1}{h_1^2} + \dfrac{\cos(\sqrt{\omega}h_2) - 1}{h_2^2}, & \omega \geq 0, \end{cases} \tag{1.21}$$

3. Let $C = -\lambda^2 < 0$. In this case D is given by

$$D(\omega) = \begin{cases} \dfrac{\cos(\sqrt{-\omega}h_1) - 1}{h_1^2} + \dfrac{\cosh(\sqrt{\gamma}h_2) - 1}{h_2^2}, & \omega \in (-\infty, 0), \\[2ex] \dfrac{\cosh(\sqrt{\omega}h_1) - 1}{h_1^2} + \dfrac{\cosh(\sqrt{\gamma}h_2) - 1}{h_2^2}, & \omega \in [0, \lambda^2), \\[2ex] \dfrac{\cosh(\sqrt{\omega}h_1) - 1}{h_1^2} + \dfrac{\cos(\sqrt{-\gamma}h_2) - 1}{h_2^2}, & \omega \in [\lambda^2, +\infty). \end{cases} \tag{1.22}$$

Thus we obtain the accurate (or exact) five-point difference equation (1.19) for the Eq. (1.1) (see, for example, [1, 2]). The function $D(\omega)$ in (1.19) can be presented as a sum of two ones, i.e.,

$$D(\omega) = D_1(\omega) + D_2(\omega), \tag{1.23}$$

where $D_1(\omega)$ and $D_2(\omega) \equiv D_2(\gamma)$ correspond to the first and second terms in (1.20), (1.21) and (1.22), respectively.

1.2.3 Accurate Finite Difference Boundary Conditions

Now we will derive accurate difference boundary conditions for (1.7), (1.8).

If $\beta_1 = 0$ in (1.7), then we have

$$U_{10} = \frac{\chi_{10}}{\alpha_1}. \tag{1.24}$$

If $\beta_1 \neq 0$, using (1.5), (1.6) in the Taylor series expansion

$$U_{11} = U_{10} + h_1 U_{10}' + \frac{h_1^2}{2!} U_{10}'' + \frac{h_1^3}{3!} U_{10}''' + \cdots, \tag{1.25}$$

we obtain

$$U_{11} = \begin{cases} \cos(\sqrt{-\omega}h_1)U_{10} + \dfrac{\sin(\sqrt{-\omega}h_1)}{\sqrt{-\omega}} U_{10}', & \omega < 0, \\[2ex] \cosh(\sqrt{\omega}h_1)U_{10} + \dfrac{\sinh(\sqrt{\omega}h_1)}{\sqrt{\omega}} U_{10}', & \omega \geq 0. \end{cases} \tag{1.26}$$

Then finding U'_{10} from (1.7) and substituting it in (1.26) we get

$$U_{11} = \theta_1(\omega)U_{10} + \theta_2(\omega), \tag{1.27}$$

where $\theta_1(\omega)$ and $\theta_2(\omega)$ are given by

$$\theta_1(\omega) = \begin{cases} \cos(\sqrt{-\omega}h_1) + \dfrac{\alpha_1}{\beta_1}\dfrac{\sin(\sqrt{-\omega}h_1)}{\sqrt{-\omega}}, & \omega < 0, \\[3mm] \cosh(\sqrt{\omega}h_1) + \dfrac{\alpha_1}{\beta_1}\dfrac{\sinh(\sqrt{\omega}h_1)}{\sqrt{\omega}}, & \omega \geq 0, \end{cases} \tag{1.28}$$

$$\theta_2(\omega) = \begin{cases} -\dfrac{\chi_{10}}{\beta_1}\dfrac{\sin(\sqrt{-\omega}h_1)}{\sqrt{-\omega}}, & \omega < 0, \\[3mm] -\dfrac{\chi_{10}}{\beta_1}\dfrac{\sinh(\sqrt{\omega}h_1)}{\sqrt{\omega}}, & \omega \geq 0. \end{cases} \tag{1.29}$$

Analogously, it is easy to verify that the exact difference boundary conditions for $U_1(x)$ at point $x = b$ are given by the expressions

$$U_{1N} = \frac{\chi_{2N}}{\alpha_2}, \quad \beta_2 = 0, \tag{1.30}$$

$$U_{1N-1} = \theta_3(\omega)U_{1N} + \theta_4(\omega), \quad \beta_2 \neq 0, \tag{1.31}$$

where $\theta_3(\omega)$ and $\theta_4(\omega)$ are given by

$$\theta_3(\omega) = \begin{cases} \cos(\sqrt{-\omega}h_1) + \dfrac{\alpha_2}{\beta_2}\dfrac{\sin(\sqrt{-\omega}h_1)}{\sqrt{-\omega}}, & \omega < 0, \\[3mm] \cosh(\sqrt{\omega}h_1) + \dfrac{\alpha_2}{\beta_2}\dfrac{\sinh(\sqrt{\omega}h_1)}{\sqrt{\omega}}, & \omega \geq 0, \end{cases} \tag{1.32}$$

$$\theta_4(\omega) = \begin{cases} -\dfrac{\chi_{2N}}{\beta_2}\dfrac{\sin(\sqrt{-\omega}h_1)}{\sqrt{-\omega}}, & \omega < 0. \\[3mm] -\dfrac{\chi_{2N}}{\beta_2}\dfrac{\sinh(\sqrt{\omega}h_1)}{\sqrt{\omega}}, & \omega \geq 0. \end{cases} \tag{1.33}$$

In the same way, as before, one can construct the accurate difference boundary conditions for $U_2(y)$. We omit the evaluation and present only the final results:

$$U_{20} = \frac{\chi_{30}}{\alpha_3}, \quad \beta_3 = 0, \tag{1.34}$$

$$U_{21} = \bar{\theta}_1(\gamma)U_{20} + \bar{\theta}_2(\gamma), \quad \beta_3 \neq 0, \tag{1.35}$$

where

$$\bar{\theta}_1(\gamma) = \begin{cases} \cos(\sqrt{-\gamma}h_2) + \dfrac{\alpha_3}{\beta_3}\dfrac{\sin(\sqrt{-\gamma}h_2)}{\sqrt{-\gamma}}, & \gamma < 0, \\[3mm] \cosh(\sqrt{\gamma}h_2) + \dfrac{\alpha_3}{\beta_3}\dfrac{\sinh(\sqrt{\gamma}h_2)}{\sqrt{\gamma}}, & \gamma \geq 0, \end{cases} \tag{1.36}$$

$$\bar{\theta}_2(\gamma) = \begin{cases} -\dfrac{\chi_{30}}{\beta_3}\dfrac{\sin(\sqrt{-\gamma}h_2)}{\sqrt{-\gamma}}, & \gamma < 0, \\[3mm] -\dfrac{\chi_{30}}{\beta_3}\dfrac{\sinh(\sqrt{\gamma}h_2)}{\sqrt{\gamma}}, & \gamma \geq 0, \end{cases} \tag{1.37}$$

and

$$U_{2M} = \frac{\chi_{4M}}{\alpha_4}, \quad \beta_4 = 0, \tag{1.38}$$

$$U_{2M-1} = \bar{\theta}_3(\gamma)U_{2M} + \bar{\theta}_4(\gamma), \quad \beta_4 \neq 0, \tag{1.39}$$

where

$$\bar{\theta}_3(\gamma) = \begin{cases} \cos(\sqrt{-\gamma}h_2) + \dfrac{\alpha_4}{\beta_4}\dfrac{\sin(\sqrt{-\gamma}h_2)}{\sqrt{-\gamma}}, & \gamma < 0, \\[3mm] \cosh(\sqrt{\gamma}h_2) + \dfrac{\alpha_4}{\beta_4}\dfrac{\sinh(\sqrt{\gamma}h_2)}{\sqrt{\gamma}}, & \gamma \geq 0, \end{cases} \tag{1.40}$$

$$\bar{\theta}_4(\gamma) = \begin{cases} -\dfrac{\chi_{4M}}{\beta_4}\dfrac{\sin(\sqrt{-\gamma}h_2)}{\sqrt{-\gamma}}, & \gamma < 0, \\[3mm] -\dfrac{\chi_{4M}}{\beta_4}\dfrac{\sinh(\sqrt{\gamma}h_2)}{\sqrt{\gamma}}, & \gamma \geq 0. \end{cases} \tag{1.41}$$

1.2.4   Method for Solving the Finite Difference Equations

In this section we consider a method for solving the finite difference equations (1.19). For this purpose we rewrite it in the from

$$U_{2j}(\Lambda_1 - 2D_1(\omega))U_{1i} + U_{1i}(\Lambda_2 - 2D_2(\gamma))U_{2j} = 0, \tag{1.42}$$
$$i = 1, \ldots, N-1, \quad j = 1, \ldots, M-1,$$

in which we have used (1.23). From this is clear, that Eq. (1.42) will be fulfilled if we choose U_{1i} and U_{2j} such that

$$(\Lambda_1 - 2D_1(\omega))U_{1i} = 0, \quad i = 1, \ldots, N-1, \tag{1.43}$$

$$(\Lambda_2 - 2D_2(\gamma))U_{2j} = 0, \quad j = 1, \ldots, M-1. \tag{1.44}$$

The last weakly coupled systems of equations (1.43), (1.44) is split into two equations with corresponding boundary conditions. First, we consider the Eq. (1.43) subject to boundary conditions (1.24), (1.27) and (1.30), (1.31). According to (1.1), (1.2) and (1.3) the function $U_1(x)$ will be determined within an arbitrary multiplicative constant. Therefore the three-point finite difference equation (1.43) can be solved by shooting method starting with U_{10}^0, U_{11}^0 and ω_0 which are required to be known. Using (1.24), (1.27), we find U_{10} or U_{11} depending on the β_1. For example, if $\beta_1 = 0$ then U_{10} is determined by (1.24) and U_{11} and ω to be chosen arbitrary. Otherwise, U_{11} is determined by (1.27) and U_{10} and ω to be chosen arbitrary. Note, that when $C \neq 0$ one of the boundary conditions (1.7), (1.8) is assumed to be homogeneous. For Laplace's equation we always can leads to equation with homogeneous boundary conditions by change of variables. The exact value of parameter ω must satisfy

$$\Phi(\omega) = 0, \tag{1.45}$$

where $\Phi(\omega)$, for examples, when $\chi_{10} = 0$ defined by

$$\Phi(\omega) = \begin{cases} U_{1N} - \dfrac{1}{\alpha_2}\chi_{2N}, & \beta_2 = 0, \\ U_{1N-1} - \theta_3(\omega)U_{1N} - \theta_4(\omega), & \beta_2 \neq 0. \end{cases} \tag{1.46}$$

The nonlinear Eq. (1.45) can be solved by Newton's method:

$$\omega_{k+1} = \omega_k - \frac{\Phi(\omega_k)}{\Phi'(\omega_k)}, \quad k = 0, 1, 2, \ldots. \tag{1.47}$$

The value $\Phi'(\omega_k)$ in the dominator of (1.47) is found be differentiating the Eqs. (1.46) and (1.43) with respect to ω. The iteration process (1.47) is terminated by criterion.

$$|\omega_{k+1} - \omega_k| \leq \varepsilon, \tag{1.48}$$

where ε is a reassigned accuracy.

If the evaluation of $\Phi'(\omega)$ causes some difficulty we can use secant method instead of Newton's ones. After finding ω the three-point difference equations (1.44) with boundary conditions (1.34), (1.35), (1.38), (1.39) can be solved by elimination method.

1.2.5   Numerical Results

We have tested the efficiency and accuracy of FDM (1.14) on the several examples.

Example 1.1 We consider the BVP

$$\left(\frac{\partial^2}{\partial x^2} + \frac{\partial^2}{\partial y^2}\right) u(x, y) = 0, \quad 0 \leq x, y \leq 1, \tag{1.49}$$

with boundary conditions

Table 1.1 Computed values of $u_{ij} = U_{1i}U_{2j}$ for $N = 5$ and $M = 4$

i, j	0	1	2	3	4	5
4	0.0000000	−0.5877853	0.9510565	−0.9510565	0.5877853	0.0000000
	(0.0000000)	(−0. 5877853)	(0.9510565)	(−0.9510565)	(0. 5877853)	(0.0000000)
3	0.0000000	−0.0052802	0.0085436	−0.0085436	0.0052802	0.0000000
	(0.0000000)	(−0.0052802)	(0.0085436)	(−0.0085436)	(0.0052802)	(0.0000000)
2	0.0000000	−0.0052802	0.0000767	−0.0000767	0.0000474	0.0000000
	(0.0000000)	(−0.0000474)	(0.0000767)	(−0.0000767)	(0.0000474)	(0.0000000)
1	0.0000000	−0.0000004	0.0000007	−0.0000007	0.0000004	0.0000000
	(0.0000000)	(−0.0000004)	(0.0000007)	(−0.0000007)	(0.0000004)	(0.0000000)
0	0.0000000	0.0000000	0.0000000	0.0000000	0.0000000	0.0000000
	(0.0000000)	(0.0000000)	(0.0000000)	(0.0000000)	(0.0000000)	(0.0000000)

$$u(0, y) - u'_x(0, y) = 6\pi \frac{\sinh(6\pi y)}{\sinh(6\pi)}, \quad u(1, y) = 0,$$
$$u(x, 0) = 0, \quad u(x, 1) = \sin(6\pi x). \tag{1.50}$$

The exact solution is given by

$$u(x, y) = \sin(6\pi x)\frac{\sinh(6\pi y)}{\sinh(6\pi)}. \tag{1.51}$$

In Table 1.1 we present the computed values of $u_{ij} = U_{1i}U_{2j}$ (the exact values of $u_{ij} = u(x_i, y_j)$ present in brackets) for $N = 5$ and $M = 4$. In order to use secant method we need two first approximations ω_0 and ω_1 to ω. The iteration was terminated by criterion (1.48) with $\varepsilon = 10^{-7}$.

Example 1.2 We consider the BVP (1.49) with boundary conditions

$$u(0, y) = u(1, y) = 0,$$
$$u(x, 0) = 0, \quad u(x, 1) = \sin(\pi x). \tag{1.52}$$

The exact solution is given by

$$u(x, y) = \sin(\pi x)\frac{\sinh(\pi y)}{\sinh(\pi)}. \tag{1.53}$$

In Table 1.2 we present the computed values of $u_{ij} = U_{1i}U_{2j}$ (the exact values of $u_{ij} = u(x_i, y_j)$ present in brackets) for $N = 6$ and $M = 6$. In order to use secant method we need two first approximations ω_0 and ω_1 to ω. In this example choose $\omega_0 = -5$ and $\omega_1 = -6$. The exact value of ω is $\omega = -\pi^2$. The convergence of ω_k was tabulated in Table 1.3. The iteration was terminated by criterion (1.48) with $\varepsilon = 10^{-7}$.

Table 1.2 Computed values of $u_{ij} = U_{1i}U_{2j}$ for $N = 6$ and $M = 6$

i, j	0	1	2	3	4	5	6
6	0.0000000	0.5000000	0.8660254	1.0000000	0.8660254	0.5000000	0.0000000
	(0.0000000)	(0. 5000000)	(0.8660254)	(1.0000000)	(0.8660254)	(0. 5000000)	(0.0000000)
5	0.0000000	0.2951674	0.5112450	0.5903348	0.5112450	0.2951674	0.0000000
	(0.0000000)	(0.2951674)	(0.5112450)	(0.5903348)	(0.5112450)	(0.2951674)	(0.0000000)
4	0.0000000	0.1731224	0.2998568	0.3462448	0.2998568	0.1731224	0.0000000
	(0.0000000)	(0.1731224)	(0.2998568)	(0.3462448)	(0.2998568)	(0.1731224)	(0.0000000)
3	0.0000000	0.0996342	0.1725715	0.1992684	0.1725715	0.0996342	0.0000000
	(0.0000000)	(0.0996342)	(0.1725715)	(0.1992684)	(0.1725715)	(0.0996342)	(0.0000000)
2	0.0000000	0.0540911	0.0936885	0.1081821	0.0936885	0.0540911	0.0000000
	(0.0000000)	(0.0540911)	(0.0936885)	(0.1081821)	(0.0936885)	(0.0540911)	(0.0000000)
1	0.0000000	0.0237192	0.0410828	0.0474384	0.0410828	0.0237192	0.0000000
	(0.0000000)	(0.0237192)	(0.0410828)	(0.0474384)	(0.0410828)	(0.0237192)	(0.0000000)
0	0.0000000	0.0000000	0.0000000	0.0000000	0.0000000	0.0000000	0.0000000
	(0.0000000)	(0.0000000)	(0.0000000)	(0.0000000)	(0.0000000)	(0.0000000)	(0.0000000)

Table 1.3 The convergence of ω_k

ω_2	ω_3	ω_4	ω_5	ω_6	ω_7
-8.4944719	-9.4982904	-9.8358860	-9.8687281	-9.8696023	-9.86960448

Table 1.4 Computed values of $u_{ij} = U_{1i}U_{2j}$ for $N = 6$ and $M = 4$

i, j	0	1	2	3	4	5	6
4	2.7182818	3.2112705	3.7936679	4.4816891	5.2944901	6.2547010	7.3890561
	(2.7182818)	(3.2112705)	(3.7936679)	(4.4816891)	(5.2944901)	(6.2547010)	(7.3890561)
3	2.1170000	2.5009400	2.9545115	3.4903430	4.1233530	4.8711660	5.7546027
	(2.1170000)	(2.5009400)	(2.9545115)	(3.4903430)	(4.1233530)	(4.8711660)	(5.7546027)
2	1.6487213	1.9477340	2.3009759	2.7182818	3.2112705	3.7936679	4.4816891
	(1.6487213)	(1.9477340)	(2.3009759)	(2.7182818)	(3.2112705)	(3.7936679)	(4.4816891)
1	1.2840254	1.5168968	1.7920018	2.1170000	2.5009400	2.9545115	3.4903430
	(1.2840254)	(1.5168968)	(1.7920018)	(2.1170000)	(2.5009400)	(2.9545115)	(3.4903430)
0	1.0000000	1.1813604	1.3956124	1.6487213	1.9477340	2.3009759	2.7182818
	(1.0000000)	(1.1813604)	(1.3956124)	(1.6487213)	(1.9477340)	(2.3009759)	(2.7182818)

Example 1.3 We consider the BVP

$$\left(\frac{\partial^2}{\partial x^2} + \frac{\partial^2}{\partial y^2}\right) u(x, y) + 2u(x, y) = 0, \quad 0 \le x, y \le 1, \tag{1.54}$$

with boundary conditions

Table 1.5 The convergence of ω_k

ω_2	ω_3	ω_4	ω_5
1.1269983	1.0086468	1.0000778	1.0000000

$$u(0, y) - u'_x(0, y) = 0, \quad u(1, y) = \exp(1 + y),$$
$$u(x, 0) = \exp(x), \quad u(x, 1) = \exp(1 + x). \tag{1.55}$$

The exact solution is given by

$$u(x, y) = \exp(x + y). \tag{1.56}$$

In Table 1.4 we present the computed the values of $u_{ij} = U_{1i}U_{2j}$ (the exact values of $u_{ij} = u(x_i, y_j)$ present in brackets) for $N = 6$ and $M = 4$. In order to use secant method we need two first approximations ω_0 and ω_1 to ω. In this example we were choose $\omega_0 = 3$ and $\omega_1 = 2$. The exact value of ω is $\omega = 1$. The convergence of ω_k was tabulated in Table 1.5. The iteration was terminated by criterion (1.48) with $\varepsilon = 10^{-7}$.

1.3   Accurate Finite Difference Method for the Wave Equation

Let $\Omega = (0, l) \times (0, +\infty)$ be an open rectangular domain in Euclidean $\mathbb{R}^2$ space with boundary given by $\partial\Omega$. The aim is to determine a function $u(x, t)$, satisfying equation [5]

$$\frac{\partial^2 u(x, t)}{\partial t^2} = c^2 \frac{\partial^2 u(x, t)}{\partial x^2}, \quad (x, t) \in \Omega, \tag{1.57}$$

with boundary conditions

$$\alpha_1 u(0, t) - \beta_1 \left.\frac{\partial u(x, t)}{\partial x}\right|_{x=0} = \xi_1(t),$$
$$\alpha_2 u(l, t) + \beta_2 \left.\frac{\partial u(x, t)}{\partial x}\right|_{x=l} = \xi_2(t), \tag{1.58}$$

and the initial conditions

$$u(x, 0) = \varphi(x), \quad \left.\frac{\partial u(x, y)}{\partial x}\right|_{t=0} = \psi(x), \tag{1.59}$$

where c, α_1, α_2, β_1 and β_2 are given numbers, and $\xi_1(t)$, $\xi_2(t)$, $\varphi(x)$ and $\psi(x)$ are given smooth functions.

It is well known that the stabilized oscillation problems and diffusing processes in gas lead to the so called wave equation (1.57) with a positive coefficient. We will assume that the problem (1.57), (1.58), (1.59) has an unique and sufficiently smooth solution. By virtue of separation of variables method looking the solution $u(x, y)$ of Eqs. (1.57), (1.58), (1.59) in the form

$$u(x, t) = U_1(x)U_2(t), \tag{1.60}$$

we arrive to equation

$$c^2 \frac{U_1''(x)}{U_1(x)} - \frac{U_2''(t)}{U_2(t)} = 0, \tag{1.61}$$

that splits into two independent equations

$$U_1''(x) = \omega U_1(x), \tag{1.62}$$
$$U_2''(t) = \gamma U_2(t), \quad \gamma = c^2\omega, \tag{1.63}$$

where the unknown separation constant ω is to be found. By virtue of (1.60) the boundary conditions (1.58) is split for $U_1(x)$ and $U_2(t)$

$$\alpha_1 U_1(0) - \beta_1 U_1'(0) = \chi_{10}, \quad \alpha_2 U_1(l) + \beta_2 U_1'(l) = \chi_{2N}, \tag{1.64}$$

and initial conditions (1.59)

$$U_2(0) = \chi_{30}, \quad U_2'(0) = \chi_{4M}. \tag{1.65}$$

The solution of the BVP (1.62), (1.64) is expressed by the formulas (1.9), (1.10) with $a = 0$, $b = l$. Analogously, we can find the solutions of the BVP (1.63) and (1.65) in closed form

$$U_2(t) = \begin{cases} \chi_{30}\cos(\sqrt{-\gamma}t) + \chi_{4M}\dfrac{\sin(\sqrt{-\gamma}t)}{\sqrt{-\gamma}}, & \gamma < 0, \\[2mm] \chi_{30}\cosh(\sqrt{\gamma}t) + \chi_{4M}\dfrac{\sinh(\sqrt{\gamma}t)}{\sqrt{\gamma}}, & \gamma \ge 0. \end{cases} \tag{1.66}$$

1.3.1 Construction of the Accurate Finite Difference Method

For the numerical solution of problem (1.57), (1.58), (1.59) is introduced the uniform rectangular grid Ω_h:

$$\Omega_h = \{(x_i, t_j) | x_i = ih_1, \quad t_j = jh_2, \quad i = 0, 1, \ldots, N, \quad j = 0, 1, \ldots, M\}, \tag{1.67}$$

where $h_1 = l/N$ and h_2 is arbitrary mesh sizes in the x and t directions, respectively. Usually, the Eq. (1.57) is approximated by the five-point difference equation

$$c^2 \frac{y_{i+1,j} - 2y_{i,j} + y_{i-1,j}}{h_1^2} - \frac{y_{i,j+1} - 2y_{i,j} + y_{i,j-1}}{h_2^2} = 0, \tag{1.68}$$
$$i = 1, \ldots, N-1, \quad j = 1, \ldots, M-1.$$

The local discretization error of the Eq. (1.68) is of $O(h_1^2 + h_2^2)$ order. Now we describe how to derive the accurate difference method for Eq. (1.57). To this end, we consider expression

$$(c^2 \Lambda_1 - \Lambda_2) u_{ij} = c^2 \frac{u_{i+1,j} - 2u_{ij} + u_{i-1,j}}{h_1^2} - \frac{u_{i,j+1} - 2u_{i,j} + u_{i,j-1}}{h_2^2}, \tag{1.69}$$

where $u_{i,j} = u(x_i, t_j)$. If we denote by U_{1i} and U_{2j} the values of $U_1(x_i)$ and $U_2(t_j)$, respectively, the using (1.60) the Eq. (1.69) maybe written as

$$(c^2 \Lambda_1 - \Lambda_2) u_{ij} = c^2 U_{2j} \Lambda_1 U_{1i} - U_{1i} \Lambda_2 U_{2j}, \tag{1.70}$$

Due to smoothness assumption of solution $u(x, t)$, as well as, functions $U_1(x)$ and $U_2(t)$, the Taylor series expansion yields

$$\Lambda_1 U_{1i} = U_1''(x_i) + 2 \sum_{k=1}^{\infty} \frac{h_1^{2k} U_1^{2k+2}(x_i)}{(2k+2)!}, \tag{1.71}$$

$$\Lambda_2 U_{2j} = U_2''(t_j) + 2 \sum_{k=1}^{\infty} \frac{h_2^{2k} U_2^{2k+2}(t_j)}{(2k+2)!}. \tag{1.72}$$

Because of (1.62), (1.63) we have

$$U_1^{(2k)} = \omega^k U_1, \quad U_2^{(2k)} = \gamma^k U_2, \quad k = 1, 2, \ldots. \tag{1.73}$$

Taking into account (1.71)–(1.73) in (1.70) it follows that

$$\left(c^2 \Lambda_1 - \Lambda_2 - 2 \sum_{k=1}^{\infty} \frac{c^2 h_1^{2k} \omega^{k+1} - h_2^{2k} \gamma^{k+1}}{(2k+2)!} \right) u_{ij} = 0, \tag{1.74}$$
$$i = 1, 2, \ldots, N-1, \quad j = 1, 2, \ldots, M-1.$$

The difference Eq. (1.74) contains unknown nonzero parameter ω and therefore it may be considered as a nonlinear equation with respect to the parameter ω and u_{ij}. The series in (1.74) may be expressed through analytical functions depending on the sign of quantities ω and β and thereby the Eq. (1.74) can be written as

$$(c^2 \Lambda_1 - \Lambda_2 - 2D(\omega)) u = 0, \quad (x, t) \in \Omega_h. \tag{1.75}$$

There are two cases

1. Let $\omega < 0$. Then it is easy to show that

$$D(\omega) = \frac{\cos(\sqrt{-\omega}h_1) - \frac{\omega h_1^2}{2} - 1}{h_1^2} - \frac{\cos(\sqrt{-\gamma}h_2) - \frac{\gamma h_2^2}{2} - 1}{h_2^2}. \qquad (1.76)$$

2. Let $\omega \geq 0$. In this case $D(\omega)$ is given by

$$D(\omega) = \frac{\cosh(\sqrt{\omega}h_1) - \frac{\omega h_1^2}{2} - 1}{h_1^2} - \frac{\cosh(\sqrt{\gamma}h_2) - \frac{\gamma h_2^2}{2} - 1}{h_2^2}. \qquad (1.77)$$

Thus we obtain the accurate (or exact) five-point difference Eq. (1.75) for the Eq. (1.57) (see, for example, [1, 2]). The function $D(\omega)$ in (1.75) can be presented as a sum of two ones, i.e.,

$$D(\omega) = D_1(\omega) - D_2(\omega), \qquad (1.78)$$

where $D_1(\omega)$ and $D_2(\omega) \equiv D_2(\gamma)$ correspond to the first and second terms in (1.76) and (1.77), respectively.

1.3.2   Accurate Finite Difference Initial and Boundary Conditions

The accurate difference boundary conditions for (1.64) are presented in Sect. 1.2.3 (see the formulas (1.26)–(1.33)).

In the same way, one can construct the accurate difference initial conditions for (1.65). We omit the evaluation and present only the final results:

$$U_{20} = \chi_{30}, \qquad (1.79)$$
$$U_{21} = \bar{\theta}_1(\gamma)\chi_{30} + \bar{\theta}_2(\gamma)\chi_{4M}, \qquad (1.80)$$

where $\bar{\theta}_1(\gamma), \bar{\theta}_2(\gamma)$ are defined by

$$\bar{\theta}_1(\gamma) = \begin{cases} \cos(\sqrt{-\gamma}h_2), & \gamma < 0, \\ \cosh(\sqrt{\gamma}h_2), & \gamma \geq 0, \end{cases} \qquad (1.81)$$

$$\bar{\theta}_2(\gamma) = \begin{cases} \dfrac{\sin(\sqrt{-\gamma}h_2)}{\sqrt{-\gamma}}, & \gamma < 0, \\ \dfrac{\sinh(\sqrt{\gamma}h_2)}{\sqrt{\gamma}}, & \gamma \geq 0. \end{cases} \qquad (1.82)$$

1.3.3   Method for Solving the Finite Difference Equations

In this section we consider a method for solving the finite difference equation (1.75). For this purpose we rewrite it in the from

$$c^2 U_{2j}(\Lambda_1 - 2D_1(\omega))U_{1i} - U_{1i}(\Lambda_2 - 2D_2(\gamma))U_{2j} = 0, \tag{1.83}$$
$$i = 1, \ldots, N - 1, \quad j = 1, \ldots, M - 1,$$

in which we have used (1.78). From this is clear, that Eq. (1.83) will be fulfilled if we choose U_{1i} and U_{2j} such that

$$(\Lambda_1 - 2D_1(\omega))U_{1i} = 0, \quad i = 1, \ldots, N - 1, \tag{1.84}$$
$$(\Lambda_2 - 2D_2(\gamma))U_{2j} = 0, \quad j = 1, \ldots, M - 1. \tag{1.85}$$

The last weakly coupled system of Eqs. (1.84), (1.85) is splitted into two equations with corresponding boundary conditions.

Numerical methods for solving the Eq. (1.84) are discussed in Sect. 1.2.4.

After finding ω the three-point difference equations (1.85) with boundary conditions (1.79)–(1.82) can be solved by elimination method.

1.3.4   Numerical Results

We have tested the efficiency and accuracy of FDM on the several examples by using the MatLab program.

Example 1.4   We consider the BVP

$$\frac{\partial^2 u(x, t)}{\partial t^2} = \frac{\partial^2 u(x, t)}{\partial x^2}, \quad 0 \le x \le 1, \quad t \ge 0, \tag{1.86}$$

with boundary conditions

$$u(0, t) = \sin(\pi t), \quad u(1, t) = -\sin(\pi t), \tag{1.87}$$

and the initial conditions

$$u(x, 0) = 0, \quad \frac{\partial u(x, 0)}{\partial t} = \pi \cos(\pi x). \tag{1.88}$$

The exact solution of this BVP is

$$u(x, t) = \cos(\pi x) \sin(\pi t). \tag{1.89}$$

Table 1.6 Computational results for **Example** 1.4

x	$t = 1/4$		$t = 1/2$	
	Exact solution	Numerical solution	Exact solution	Numerical solution
0.1	0.672499	0.672499	0.951057	0.951057
0.2	0.572061	0.572061	0.809017	0.809017
0.3	0.415627	0.415627	0.587785	0.587785
0.4	0.218508	0.218508	0.309017	0.309017
0.5	0.000000	0.000000	0.000000	0.000000
0.6	−0.218508	−0.218508	−0.309017	−0.309017
0.7	−0.415627	−0.415627	−0.587785	−0.587785
0.8	−0.572061	−0.572061	−0.809017	−0.809017
0.9	−0.672499	−0.672499	−0.951057	−0.951057

Example 1.5 We consider the BVP (1.86) with boundary conditions

$$u(0, t) = \cos(\pi t), \quad u(1, t) = 0, \tag{1.90}$$

and the initial conditions

$$u(x, 0) = \cos(\pi x), \quad \frac{\partial u(x, 0)}{\partial t} = 0. \tag{1.91}$$

The exact solution of this BVP is

$$u(x, t) = \frac{1}{2} \left(\cos(\pi(x + t)) + \cos(\pi(x - t)) \right). \tag{1.92}$$

We compare the results obtained of the **Examples** 1.4 and 1.5 by the procedure in previous section with the accurate FDM introduced in Tables 1.6 and 1.7, respectively.

1.4   Statement of Problem and Exact Finite Difference Method for Helmholtz Equation

We consider the following Helmholtz equation

$$\left(\frac{\partial^2}{\partial x^2} + \frac{\partial^2}{\partial y^2} \right) u(x, y) + k^2 u(x, y) = 0 \quad \text{in} \quad \Omega, \tag{1.93}$$

where k is a fixed wavenumber and $u(x, y)$ is solution subjected to Direchlet or mixed boundary conditions on $\partial\Omega$. We assume that the basic conditions for the existence of a

Table 1.7 Computational results for **Example** 1.5

x	$t = 1/5$		$t = 1/10$	
	Exact solution	Numerical solution	Exact solution	Numerical solution
0.1	0.769421	0.769421	0.904508	0.904508
0.2	0.654508	0.654508	0.769421	0.769421
0.3	0.475528	0.475528	0.559017	0.559017
0.4	0.250000	0.250000	0.293893	0.293893
0.5	0.555112	0.555112	0.555112	0.555112
0.6	−0.250000	−0.250000	−0.293893	−0.293893
0.7	−0.475528	−0.475528	−0.559017	−0.559017
0.8	−0.654508	−0.654508	−0.769421	−0.769421
0.9	−0.769421	−0.769421	−0.904508	−0.904508

unique solution are satisfied. In order to solve problem (1.93) numerically it is necessary to introduce a discrete grid on $\bar{\Omega} = \Omega \cup \partial\Omega$. Let S_h denote a uniform rectangular region $\bar{\Omega}$ such that for some integers N and M,

$$S_h = \{(x_i, y_j) | x_i = x_0 + ih, \ y_j = y_0 + jh, \ i = 0, 1, \ldots, N \ j = 0, 1, \ldots, M\}, \quad (1.94)$$

where h is the mesh size of grid.

Let U_{ij} be a grid function that satisfies the following difference equation for $i = 1, \ldots, N - 1, j = 1, \ldots, M - 1$

$$U_{ij} = \frac{1}{4A(kh)} \left(U_{i-1,j} + U_{i+1,j} + U_{i,j-1} + U_{i,j+1} \right). \quad (1.95)$$

Here $A(kh)$ is determined in diverse way depending on choosing methods. For example:

$$A(kh) = \begin{cases} T_0(kh) = 1 - \dfrac{(kh)^2}{4}, & (1.96a) \\[2mm] J_0(kh), & (1.96b) \\[2mm] C(kh). & (1.96c) \end{cases}$$

The case (1.96a) corresponds to a standard FDM. The case (1.96b) with Bessel function of order zero

$$J_0(x) = \sum_{m=0}^{\infty} \frac{(-x^2)^m}{2^{2m}(m!)^2}, \quad (1.97)$$

was proposed in [8]. In [4] was proposed the exact FDM for (1.93) and has a form (1.95) with

$$C(kh) = 1 - \frac{k^2 h^2}{4} + \frac{1}{2}\left(\sum_{m=2}^{\infty}\frac{\omega^m h^{2m}}{(2m)!} + \sum_{m=2}^{\infty}\frac{\gamma^m h^{2m}}{(2m)!}\right),\qquad(1.98)$$

where ω and γ are a separation constants in $u(x, y) = v(x)z(y)$:

$$\frac{v''(x)}{v(x)} = \omega,\qquad(1.99)$$

$$\frac{z''(y)}{z(y)} = \gamma,\quad \gamma = -k^2 - \omega.\qquad(1.100)$$

It is easy to show that the Eq. (1.93) has a planar wave solution

$$u(x, y) = \exp\big(\pm\iota(k_1 x + k_2 y)\big)\qquad(1.101)$$

with

$$k_1 = k\cos(\theta),\quad k_2 = k\sin(\theta),\quad \theta \in [0, \pi].\qquad(1.102)$$

From (1.101) it clear that the solution $u(x, y)$ has a form of products of two functions and according to (1.102) we have

$$\omega = -k_1^2,\quad \gamma = -k^2 - \omega = -k^2 + k_1^2 = -k_2^2.\qquad(1.103)$$

If we take into account (1.103), then the expression (1.98) has a form

$$C(kh) \equiv C(k_1 h, k_2 h) = 1 - \frac{k_1^2 h^2 + k_2^2 h^2}{4}$$

$$+ \frac{1}{2}\left(\sum_{m=2}^{\infty}\frac{(-1)^m k_1^{2m} h^{2m}}{(2m)!} + \sum_{m=0}^{\infty}\frac{(-1)^m k_2^{2m} h^{2m}}{(2m)!}\right)$$

$$= \frac{1}{2}(\cos(k_1 h) + \cos(k_2 h)).\qquad(1.104)$$

Thus, the FDM (1.95) with (1.96c) is exact one. Moreover, the exactness of the method (1.95), (1.96c) is verified by a direct computations (see also (1.95) in [8, 9]). The averaged value of Eq. (1.104) over all angles θ equal to $J_0(kh)$.

According to (1.96a), (1.96b) and (1.96c) we have

$$T_0(kh) - C(k_1 h, k_2 h) = O((kh)^4),\qquad(1.105)$$

$$J_0(kh) - C(k_1 h, k_2 h) = O((kh)^4).\qquad(1.106)$$

It means that the methods (1.95), (1.96a) and (1.95), (1.96b) have a second-order accuracy, i.e.

$$\|u - U\|_\infty = O((kh)^2). \tag{1.107}$$

Remark 1.1 Note that if $\lambda^2 = -k^2 > 0$ and/or $\lambda_1^2 = -k_1^2 > 0$ and/or $\lambda_2^2 = -k_2^2 > 0$ in (1.96b), (1.96c) the Bessel function $J_0(kh)$ replaced by the modified Bessel function $I_0(\lambda h) = J_0(kh)$ of order zero [10, 11] and/or the trigonometrical $\cos(k_j h)$ function replaced by hyperbolic $\cosh(\lambda_j h)$ function, respectively.

In [10] was pointed out that the method (1.95), (1.96b) has a higher accuracy (in case of $\lambda^2 = -k^2 > 0$, i.e., $J_0(kh) = I_0(\lambda h)$), but not given theoretical analysis for accuracy of this method. But for some angles of incident θ, the accuracy of method (1.95), (1.96b) may be increased as shown the following.

Theorem 1.1 *Let $u \in C^8(\Omega)$ and U_{ij} is a grid function satisfying (1.95), (1.96b), and $k_1^2 > 0$, $k_2^2 > 0$. Then holds*

$$\|u - U\|_\infty = O((kh)^6). \tag{1.108}$$

in the wave directions $\theta = \pi/8, 3\pi/8, 5\pi/8, 7\pi/8 \in [0, \pi]$.

Proof From the Taylor expansions of (1.96c) and (1.102) we get

$$C(k_1 h, k_2 h) = 1 - \frac{k^2 h^2}{4} + \frac{1}{2} \sum_{m=2}^{\infty} \frac{(-1)^m (kh)^{2m}}{(2m)!} \left(\cos^{2m}(\theta) + \sin^{2m}(\theta) \right). \tag{1.109}$$

Using (1.97) and above expression, we have

$$J_0(kh) - C(k_1 h, k_2 h) = 1 - \frac{(kh)^2}{4} + \frac{(kh)^4}{64} - \frac{(kh)^6}{2304} + \cdots - \left(1 - \frac{(kh)^2}{4}\right.$$
$$+ \frac{(kh)^4}{48} \left(\cos^4(\theta) + \sin^4(\theta) \right) - \frac{(kh)^6}{1440} \left(\cos^6(\theta) + \sin^6(\theta) \right) + \cdots \right)$$
$$= -\frac{(kh)^4}{192} \cos(4\theta) + \frac{(kh)^6}{3840} \cos(4\theta) + O((kh)^8). \tag{1.110}$$

in which we have used $\cos^4\theta + \sin^4\theta = (3 + \cos(4\theta))/4$, $\cos^6\theta + \sin^6\theta = (5 + 3\cos(4\theta))/8$. From this it clear that

$$J_0(kh) - C(k_1 h, k_2 h) = O((kh)^8), \tag{1.111}$$

under conditions $\theta = \pi/8, 3\pi/8, 5\pi/8, 7\pi/8$.

Now we consider the truncation errors of method (1.95), (1.96b)

$$\psi_{ij} = U_{ij} - \frac{1}{4J_0(kh)}\Big(U_{i-1,j} + U_{i+1,j} + U_{i,j-1} + U_{i,j+1}\Big). \qquad (1.112)$$

Using (1.111), we obtain

$$\psi_{ij} = U_{ij}$$
$$- \frac{1}{4C(k_1h, k_2h)}\Big(U_{i-1,j} + U_{i+1,j} + U_{i,j-1} + U_{i,j+1}\Big) + O((kh)^8), \quad (1.113)$$

because of $\theta = \pi/8,\ 3\pi/8,\ 5\pi/8,\ 7\pi/8$. Following the estimation (1.97) in [10] we get

$$\|u - U\|_\infty \le \alpha(h,k)\max|\psi_{ij}| = O((kh)^6),$$

because of $\alpha(h,k) = O((kh)^{-2})$ and (1.113).

In [10] was pointed out that the accuracy of method (1.95), (1.96b) is better than that of the standard FDM (1.95), (1.96a) and than that indicated by the theoretical evaluation. The Theorem 1 conforms this conclusion.

As above, it is easy to show that

$$I_0(kh) - C(k_1h, k_2h)$$
$$= -\frac{(kh)^4}{192}\cos(4\theta) - \frac{(kh)^6}{3840}\cos(4\theta) + O((kh)^8) = O((kh)^8), \qquad (1.114)$$

in the directions $\theta = \pi/8,\ 3\pi/8,\ 5\pi/8,\ 7\pi/8$, where $k^2 = k_1^2 + k_2^2,\ k_1^2 > 0$ and $k_2^2 > 0$.

The exact FDM (1.95), (1.96c) is easily extended to the three-dimensional Helmholtz equation and has a form

$$U_{i,j,l} = \frac{1}{6C(kh)}\Big(U_{i-1,j,l} + U_{i+1,j,l} + U_{i,j-1,l}$$
$$+ U_{i,j+1,l} + U_{i,j,l-1} + U_{i,j,l+1}\Big), \qquad (1.115)$$

with

$$C(kh) = C(k_1h, k_2h, k_3h) = \frac{\cos(k_1h) + \cos(k_2h) + \cos(k_3h)}{3}, \qquad (1.116)$$
$$k_1 = k\sin(\theta)\cos(\phi), \quad k_2 = k\sin(\theta)\sin(\phi),$$
$$k_3 = k\cos(\theta), \quad \theta \in [0, \pi), \quad \phi \in [0, 2\pi).$$

After average the (1.116) over all angles θ and ϕ, we have

$$\frac{1}{4\pi} \int_0^\pi d\theta \, \sin(\theta) \int_0^{2\pi} d\pi \, C(k_1 h, k_2 h, k_3 h)$$

$$= \frac{1}{4\pi} \int_0^\pi d\theta \, \sin(\theta) \int_0^{2\pi} d\pi \, \cos(kh \cos(\theta))$$

$$= \frac{\sin(kh)}{kh} \equiv j_0(kh), \tag{1.117}$$

where $j_0(x)$ is spherical Bessel function of order zero. Since

$$j_0(x) = 1 - \frac{x^2}{6} + \frac{x^4}{120} + \cdots, \tag{1.118}$$

we obtain

$$j_0(kh) - C(k_1 h, k_2 h, k_3 h) = O((kh)^4). \tag{1.119}$$

It is easy to construct the exact FDM for m-dimensional Helmholtz equation and has the same kind form as (1.115) with denominator function $C(kh)$

$$U_{i,j,\ldots,l} = \frac{1}{2mC(kh)} \sum_{z \in I_h} U(z), \tag{1.120}$$

where I_h is the neighboring set points of given stencil and

$$C(kh) = \frac{1}{m} \sum_{i=1}^m \cos(k_i h), \quad \sum_{i=1}^m k_i^2 = k^2. \tag{1.121}$$

Using the m-dimensional hyperspherical representation of k_i, $i = 1, \ldots, m$, and known analytical formulas [12]

$$\int_0^\pi d\theta \, \sin^{2\mu}(\theta) \cos(z \cos(\theta)) = \sqrt{\pi} \left(\frac{2}{z}\right)^\mu \Gamma\left(\mu + \frac{1}{2}\right) J_\mu(z), \quad \Re\mu > -\frac{1}{2},$$

$$\int_0^\pi d\theta \, \sin^{\mu-1}(\theta) = 2 \int_0^{\pi/2} d\theta \, \sin^{\mu-1}(\theta) = 2^{\mu-1} \frac{\Gamma\left(\frac{\mu}{2}\right) \Gamma\left(\frac{\mu}{2}\right)}{\Gamma(\mu)}, \quad \Re\mu > 1, \tag{1.122}$$

$$\Gamma(2z) = \frac{1}{\sqrt{\pi}} 2^{2z-1} \Gamma(z) \Gamma\left(z + \frac{1}{2}\right),$$

the averaged value of $C(kh)$ over the $(m-1)$ hyperspherical angles has the form

$$\frac{\int_{S^{m-1}} d^{m-1}V\, C(kh)}{\int_{S^{m-1}} d^{m-1}V} = \frac{\int_0^\pi d\theta_1 \sin^{m-2}(\theta_1) \cos(kh\cos(\theta_1))}{\int_0^\pi d\theta_1 \sin^{m-2}(\theta_1)}$$

$$= \Gamma\left(\frac{m}{2}\right)\left(\frac{2}{kh}\right)^{m/2-1} J_{m/2-1}(kh)$$

$$= \begin{cases} (2s-2)!!\,(kh)^{-s+1}\, J_{s-1}(kh), & m = 2s, \\ (2s-1)!!\,(kh)^{-s+1}\, j_{s-1}(kh), & m = 2s+1, \end{cases} \quad s \geq 1. \qquad (1.123)$$

Here $d^{m-1}V$ is the hyperspherical angular volume element

$$d^{m-1}V = \sin^{m-2}(\theta_1)\sin^{m-3}(\theta_2)\cdots\sin(\theta_{m-2})d\theta_1 d\theta_2 \cdots d\theta_{m-1}, \qquad (1.124)$$

$J_{s-1}(x)$ and j_{s-1} are Bessel and spherical Bessel functions, respectively. Since

$$\left(\frac{2}{x}\right)^{m/2-1}\Gamma\left(\frac{m}{2}\right) J_{m/2-1}(x) = 1 - \frac{x^2}{2m} + \frac{x^4}{8m(m+2)} + \cdots, \qquad (1.125)$$

we obtain

$$\left(\frac{2}{kh}\right)^{m/2-1}\Gamma\left(\frac{m}{2}\right) J_{m/2-1}(kh) - C(kh) = O((kh)^4). \qquad (1.126)$$

Remark 1.2 Note that if $\lambda^2 = -k^2 > 0$ the Bessel function $J_{s-1}(kh)$ and spherical Bessel function $j_{s-1}(kh)$ replaced by the modified Bessel function $I_{s-1}(\lambda h)$ and modified spherical Bessel function $i_{s-1}(\lambda h)$ with factor $\imath^{s-1}$, respectively.

1.4.1   Solutions of Discrete Equations and Calculation Techniques

Well known that to ensure an accurate numerical solution at high wave numbers of method (1.95) with (1.96a), (1.96b), we have to enforce the condition $k^2 h < 1$. However, this would imply that the number of the discretized equations is proportional to h^{-3} or k^3. This leads to an extremely large system of equations and many iterative techniques such as the conjugate gradient and multigrid methods are not capable of solving the indefinite systems. Since no discretization error is introduced in (1.95), (1.96c) one would expect that it will produce exact numerical solution for all wavenumbers even if kh and $k^2 h$ is not small.

Let

$$u(x, y) = v(x)z(y). \qquad (1.127)$$

Then (1.127) is a solution of Eq. (1.95) if functions $v(x)$, $w(y)$ satisfy the following exact finite difference equations

$$v_{i-1} - 2\cos(\sqrt{-\omega}h)v_i + v_{i+1} = 0, \quad i = 1, 2, \ldots, N-1, \tag{1.128}$$

$$z_{j-1} - 2\cos(\sqrt{-\gamma}h)z_j + z_{j+1} = 0, \quad j = 1, 2, \ldots, M-1, \tag{1.129}$$

respectively. The Eqs. (1.128) and (1.129) together with boundary conditions form a closed system.

We consider two-dimensional Helmholtz equation Eq. (1.93) with mixed boundary conditions, in which a Dirichlet are given on two-sides of the rectangular domain and Sommerfeld's radiation conditions are imposed on the remaining boundaries as in [8]. For examples, we consider the following

$$u(0, y) = f_1(y), \quad \frac{\partial u(x, y)}{\partial x}\bigg|_{x=1} = \iota k_1 u(1, y), \tag{1.130}$$

$$u(x, 0) = f_2(x), \quad \frac{\partial u(x, y)}{\partial y}\bigg|_{y=1} = \iota k_2 u(x, 1). \tag{1.131}$$

In this case taking into account Eq. (1.127) we have following boundary conditions for the fundamental solutions $v(x)$ and $z(y)$ of Eq. (1.95)

$$v(0) = \xi, \tag{1.132}$$

$$\frac{\partial v(x)}{\partial x}\bigg|_{x=1} = \iota k_1 v(1), \tag{1.133}$$

$$w(0) = \eta, \tag{1.134}$$

$$\frac{\partial z(y)}{\partial y}\bigg|_{y=1} = \iota k_2 z(1), \tag{1.135}$$

where ξ and η are given constants. Using Taylor expansions of solutions $v(x \pm h)$ and $z(y \pm h)$ in vicinity of the boundary points $x = x_N$, $y = y_M$ and the conditions (1.133), (1.135), we arrive at the exact formulas

$$\frac{\partial v(x)}{\partial x}\bigg|_{x=x_N} = \frac{v_{N+1} - v_{N-1}}{2\sin(k_1 h)/k_1}, \quad \frac{\partial z(y)}{\partial y}\bigg|_{y=y_M} = \frac{z_{M+1} - z_{M-1}}{2\sin(k_2 h)/k_2}. \tag{1.136}$$

Thus, we have the required perfect boundary conditions

$$v_{N+1} - 2\iota \sin(k_1 h)v_N - v_{N-1} = 0, \tag{1.137}$$

$$z_{M+1} - 2\iota \sin(k_2 h)z_M - z_{M-1} = 0. \tag{1.138}$$

Upon eliminating the fictitious nodal value v_{N+1} from Eq. (1.128), we can obtain a relation between v_{N-1} and v_N as

$$v_{N-1} + (\iota \sin(k_1 h) - \cos(k_1 h))v_N = 0. \tag{1.139}$$

In a similar way we have a relation between z_{M-1} and z_M as

$$z_{M-1} + (\iota \sin(k_2 h) - \cos(k_2 h))z_M = 0. \tag{1.140}$$

So, our numerical algorithm for solving Eq. (1.95) can be written in three steps as follows:
Step 1. We consider the following eigenvalue problem

$$v_{i-1} - 2\mu v_i + v_{i+1} = 0, \quad i = 1, 2, \ldots, N-1,$$

$$v_0 = \xi, \tag{1.141}$$

$$v_{N-1} + (\iota \sin(k_1 h) - \mu)v_N = 0,$$

where $\mu = \cos(\sqrt{-\omega}h)$ and $\mathbf{v} = (v_1, \ldots, v_N)^T$ are unknown eigenvalue and eigenvectors, respectively. If $\xi = 0$ (or $f_1(y) = 0$), the homogeneous eigenvalue problem (1.141) can be solved using the continuous analogue of Newton's method (see details in [13, 14] and Appendix A) with an additional normalization condition

$$(\mathbf{v}, \mathbf{v}) = \sum_{n=1}^{N} v_n^2 = 1. \tag{1.142}$$

Let $\xi = 1$ (or $f_1(y) \neq 0$). In this case the problem (1.141) has a solution, but it is not unique, and we can't use additional condition (1.142). Using Taylor expansions of solutions $v(x - h)$ in vicinity of the boundary point $x = x_N$, Eq. (1.99) and boundary condition (1.133) we have

$$v(x_{N-1}) = v(x_N) \sum_{j=0} (-1)^j \frac{\omega^j h^{2j}}{(2j)!} + v'(x_N) \sum_{j=0} (-1)^{j+1} \frac{\omega^j h^{2j+1}}{(2j+1)!}$$

$$= v(x_N) \cos(\sqrt{-\omega}h) - v'(x_N) \frac{\sin(\sqrt{-\omega}h)}{k_1}$$

$$= \left(\cos(\sqrt{-\omega}h) - \iota k_1 \frac{\sin(\sqrt{-\omega}h)}{\sqrt{-\omega}} \right) v(x_N), \tag{1.143}$$

or

$$v_{N-1} + \left(\iota k_1 \frac{\sin(\sqrt{-\omega}h)}{\sqrt{-\omega}} - \mu \right) v_N = 0. \tag{1.144}$$

Thus, problem (1.141) with an additional condition (1.144) has now a unique solution. Subtracting (1.144) from the last formula of (1.141) we obtain

$$\iota k_1 \left(\frac{\sin(\sqrt{-\omega}h)}{\sqrt{-\omega}} - \frac{\sin(k_1 h)}{k_1} \right) v_N = 0. \tag{1.145}$$

From this we have the condition $\omega = -k_1^2$ for any $v_N \neq 0$. It means that, the eigenvalue problem (1.141) leads to the system of algebraic equations with respect to the unknown $\mathbf{v} = (v_1, \ldots, v_N)^T$. For solving this problem we use the shooting method.

Step 2. Now we consider following boundary problem:

$$z_{j-1} - 2\cos(\sqrt{-\gamma}h)z_j + z_{j+1} = 0, \quad j = 1, 2, \ldots, M-1,$$

$$z_0 = \eta, \tag{1.146}$$

$$z_{M-1} + (\iota \sin(k_2 h) - \cos(\sqrt{-\gamma}h))z_M = 0.$$

Here $\mathbf{z} = (z_1, \ldots, z_M)^T$ is an unknown vector. Since we find ω in the **Step 1**, the coefficient $\gamma = -k^2 - \omega$ is known. Therefore, the boundary value problem (1.146) is solved by elimination method. Moreover, it is easy to show that, $z_j = \exp(\iota jh\sqrt{-\gamma})$ satisfy (1.146) when $\eta = 1$ and $\gamma = -k_2^2$.

Step 3. Finally we construct the solution $u(x, y)$ on the grids using calculated values of v_i and z_j

$$U_{ij} = v_i z_j. \tag{1.147}$$

The system (1.141) is a second-order difference equation with constant coefficients. Therefore, one can seeking for the solution in the from

$$v_i = \exp(\iota ihk_1) \neq 0. \tag{1.148}$$

Substituting (1.148) into (1.141) and removing the common factor we get

$$\cos(\sqrt{-\omega}h) = \cos(k_1 h) \quad \text{or} \quad \omega = -k_1^2. \tag{1.149}$$

It is easy to check that (1.148) with $\omega = -k_1^2$ satisfies boundary conditions in (1.141). In a similar way we obtain

$$z_j = \exp(\iota jhk_2). \tag{1.150}$$

Thus, the solution of system (1.139) is given by

$$U_{ij} = \exp(\iota h(ik_1 + jk_2)), \tag{1.151}$$

that coincides with values of exact solution. It means that we have exact values of solution if the incident angle θ is known. Otherwise, the system (1.141) is solved by shooting method, where the role of shooting parameter plays the incident angle θ.

1.4.2 Numerical Results

To demonstrate the effectiveness of our exact FDM (1.95), (1.96c), we carry out the following numerical simulations for two problems. To analyze the convergence of the this method, we used the maximum absolute errors of the solutions $v(x)$, $z(y)$ and $u(x, y)$:

$$\|e_v\|_{\infty,h} = \max_{0 \le i \le N} |v_i - v(x_i)|,$$

$$\|e_w\|_{\infty,h} = \max_{0 \le j \le M} |z_j - z(y_j)|, \tag{1.152}$$

$$\|e_u\|_{\infty,h} = \max_{0 \le i \le N, 0 \le j \le M} |U_{ij} - u(x_i, y_j)|.$$

The order (or Runge coefficient) of convergence of the two-dimensional method (1.95) with (1.96a) and (1.96b) is defined by the double-crowding grids

$$\text{Runge} = \frac{\|e_u\|_{\infty,h}}{\|e_u\|_{\infty,h/2}}. \tag{1.153}$$

All numerical calculations were performed using the Maple 18 system and quadruple-precision arithmetic on Intel FORTRAN Compiler.

Example 1.6

$$\left(\frac{\partial^2}{\partial x^2} + \frac{\partial^2}{\partial y^2} \right) u(x, y) = \lambda^2 u(x, y),$$

$$u(0, y) = u(1, y) = 0, \tag{1.154}$$

$$u(x, 0) = \sin(\pi x), \quad u(x, 1) = 2 \exp(-\lambda) \sin(\pi x).$$

The exact solution of this BVP is given by [15]

$$u(x, y) = \frac{2 \exp(-\lambda) \sinh(\sigma y) + \sinh(\sigma(1 - y))}{\sinh(\sigma)} \sin(\pi x), \quad \sigma^2 = \pi^2 + \lambda^2. \tag{1.155}$$

The eigenvalue problem (1.141) with homogeneous Dirichlet conditions is calculated by the Newton method. The comparison of the convergence of the Newton method with optimal parameter τ_k^{opt} (A.16), τ_k^{kal} (A.17), and $\tau_k = 1$ versus the step-size h is presented in Table 1.8. The initial values chosen that

$$v_i^0 = ih(1 - ih)^2, \quad i = 1, \dots, N - 1, \quad \omega^0 = -3.3^2, \tag{1.156}$$

corresponds to the exact $v(x) = \sin(\pi x)$ with $\omega = -k_1^2 = -\pi^2$. From Table 1.8 can see that the convergence of the Newton method with optimal parameter τ_k^{opt} (A.16) more rapidly than others. The maximum absolute errors versus the number of grid points N, M and parameter λ are shown in Table 1.9.

Example 1.7 (**Propagation of plane waves**). We now consider a two-dimensional Helmholtz equation on a unit square $\Omega = (0, 1) \times (0, 1)$ with radiation boundary conditions on two sides and the Dirichlet boundary conditions on the remaining boundaries [8]. This problem is formulated as:

Table 1.8 The comparison of the convergence of the Newton method with optimal parameter τ_n^{opt} (A.16), τ_n^{kal} (A.17), and $\tau_n = 1$. The first and second columns show the number of grid partitions N, and number of iteration n. The third and sixth columns display the values of the optimal parameter τ_n^{opt}, τ_n^{kal}, respectively. The fourth, seventh and ninth columns present corresponding the maximum absolute error $\|e_u\|_{\infty,h}$ for **Example** 1.6, while fifth, eight and tenth columns display the error of the eigenvalue $\omega^k - \omega^*$. Here $\omega^* = -\pi^2$. The factor x in the brackets denotes 10^x

N	n	τ_n^{opt}			τ_n^{kal}			$\tau_n = 1$	
		τ_n	$\|e_v\|_{\infty,h}$	$\omega^n - \omega^*$	τ_n	$\|e_v\|_{\infty,h}$	$\omega^n - \omega^*$	$\|e_v\|_{\infty,h}$	$\omega^n - \omega^*$
4	0		9.17(−02)	−1.02(−00)		9.17(−02)	−1.02(−00)	9.17(−02)	−1.02(−00)
	1	0.10530	8.23(−02)	6.63(−01)	0.10000	8.24(−02)	5.79(−01)	2.41(−00)	1.38(+01)
	2	0.57507	4.22(−02)	−4.15(−01)	0.47616	4.51(−02)	−2.41(−01)	5.93(−01)	6.33(−00)
	3	0.97286	6.76(−04)	9.99(−03)	0.99763	2.14(−03)	3.70(−02)	1.26(−01)	1.98(−00)
	4	1.00027	5.03(−08)	−4.52(−07)	0.99991	1.95(−05)	3.48(−04)	1.11(−02)	1.95(−01)
	5	0.99999	2.88(−16)	−4.25(−15)	1.00000	8.80(−10)	1.56(−08)	6.44(−05)	1.14(−03)
	6	1.00000	7.45(−32)	−7.35(−31)	1.00000	8.90(−19)	1.58(−17)	1.11(−09)	1.98(−08)
10	0		3.69(−02)	−1.02(−00)		3.69(−02)	−1.02(−00)	3.69(−02)	−1.02(−00)
	1	0.35862	2.55(−02)	1.39(−00)	0.10487	3.28(−02)	−3.12(−01)	1.07(−01)	5.69(−00)
	2	0.97106	2.33(−03)	−1.82(−01)	0.94640	6.98(−03)	5.45(−01)	2.48(−02)	2.07(−00)
	3	1.00210	1.52(−05)	−9.87(−04)	0.97627	1.20(−03)	1.17(−01)	3.37(−03)	3.36(−01)
	4	0.99999	4.40(−10)	−3.43(−08)	0.99883	4.22(−05)	4.29(−03)	6.18(−05)	6.28(−03)
	5	1.00000	5.58(−19)	−3.64(−17)	0.99999	3.02(−08)	3.07(−06)	1.12(−08)	1.14(−06)
	6	1.00000	6.23(−37)	−4.86(−35)	1.00000	7.75(−15)	7.88(−13)	1.84(−16)	1.87(−14)

Table 1.9 The convergence $\|e_w\|_{\infty,h}$ and $\|e_u\|_{\infty,h}$ versus the number of grids partitions N, M and parameter λ. The factor x in the brackets denotes 10^x

$N = M$	λ	$\|e_w\|_{\infty,h}$	$\|e_u\|_{\infty,h}$
4	10	6.38(−34)	5.15(−33)
	30	1.61(−36)	3.54(−35)
	50	6.67(−39)	2.39(−37)
	70	3.23(−41)	1.61(−39)
10	10	8.13(−38)	6.21(−37)
	30	3.94(−39)	8.07(−38)
	50	3.23(−40)	1.08(−38)
	70	3.14(−41)	1.45(−39)

Table 1.10 The convergence of the shooting method for **Example** 1.7. The first and second columns show the numbers of grid partitions N and M, and the value of k. The third–seventh columns present absolute errors $|-k_1^2 - \omega_n|$, $|\theta^* - \theta_n|$, and the maximum absolute errors $\|e_v\|_{\infty,h}$, $\|e_w\|_{\infty,h}$, $\|e_u\|_{\infty,h}$, respectively. The last column displays the total iteration number for the calculation of function v_i under condition $|\omega_{n+1} - \omega_n| < 10^{-19}$. Here $k_1 = k\cos(\theta)$ with $\theta = \pi/4$. The factor x in the brackets denotes 10^x

$N = M$	k	$\lvert -k_1^2 - \omega_n \rvert$	$\lvert \theta^* - \theta_n \rvert$	$\|e_v\|_{\infty,h}$	$\|e_w\|_{\infty,h}$	$\|e_u\|_{\infty,h}$	Iteration
4	$5\sqrt{2}$	8.70(−33)	2.65(−21)	1.14(−20)	1.14(−20)	6.38(−21)	5
	$10\sqrt{2}$	1.32(−32)	2.68(−22)	1.76(−21)	1.76(−21)	1.13(−21)	7
	$15\sqrt{2}$	2.87(−24)	2.62(−17)	3.20(−16)	3.20(−16)	2.16(−16)	6
	$20\sqrt{2}$	5.73(−36)	1.40(−24)	2.01(−23)	2.01(−23)	1.43(−23)	7
	$25\sqrt{2}$	4.92(−29)	4.03(−21)	7.59(−20)	7.59(−20)	5.06(−20)	10
	$30\sqrt{2}$	5.45(−24)	1.59(−17)	3.65(−16)	3.65(−16)	2.37(−16)	6
10	$5\sqrt{2}$	5.97(−37)	3.28(−24)	1.72(−23)	1.72(−23)	1.40(−23)	6
	$10\sqrt{2}$	1.48(−29)	1.92(−20)	1.73(−19)	1.73(−19)	1.70(−19)	7
	$15\sqrt{2}$	1.63(−28)	4.86(−20)	6.85(−19)	6.85(−19)	5.64(−19)	7
	$20\sqrt{2}$	2.01(−28)	8.51(−20)	1.46(−18)	1.46(−18)	1.37(−18)	6
	$25\sqrt{2}$	3.30(−31)	2.33(−22)	5.16(−21)	5.16(−21)	4.45(−21)	9
	$30\sqrt{2}$	1.64(−33)	1.20(−23)	3.26(−22)	3.26(−22)	2.91(−22)	8

$$\left(\frac{\partial^2}{\partial x^2} + \frac{\partial^2}{\partial y^2}\right) u(x, y) = -k^2 u(x, y),$$

$$u(0, y) = f_1(y), \qquad \frac{\partial u(x, y)}{\partial x}\bigg|_{x=1} = \imath k_1 u(1, y), \qquad (1.157)$$

$$u(x, 0) = f_2(x), \qquad \frac{\partial u(x, y)}{\partial y}\bigg|_{y=1} = \imath k_2 u(x, 1).$$

The exact solution $u(x, y) = \exp(\imath k_1 x + \imath k_2 y)$, where $(k_1, k_2) = (k \cos \theta, k \sin \theta)$ and $f_1(y)$ and $f_2(x)$ are determined such that the exact solution is a given plane wave, i.e. $f_1(y) = \exp(\imath k_2 y)$ and $f_2(x) = \exp(\imath k_1 x)$.

From the beginning, we seek $u(x, y)$ in the separated form (1.127).

The inhomogeneous eigenvalue problem (1.141) with (1.132)–(1.135), (1.144) is calculated by the shooting method. The absolute errors $| - k_1^2 - \omega_n|$, $|\theta^* - \theta_n|$, and the maximum absolute errors $\|e_v\|_{\infty,h}$, $\|e_w\|_{\infty,h}$, $\|e_u\|_{\infty,h}$ versus the number of grids N and M, and the value of k are presented in Table 1.10. The last column displays the total iteration number for the calculation of function v_i under condition $|\omega_{n+1} - \omega_n| < 10^{-19}$. Here we choose that $k_1 = k \cos(\theta)$ with $\theta = \pi/4$.

We also directly solved the two-dimensional method (1.95) with $T_0(kh)$ (1.96a), $J_0(kh)$ (1.96b). Mixed boundary conditions of Eq. (1.157) replaced by (see Eqs. (1.137) and (1.138))

$$U_{N+1,j} - 2\imath \sin(k_1 h) U_{N,j} - U_{N-1,j} = 0, \quad j = 0, \dots, M, \qquad (1.158)$$

$$U_{i,M+1} - 2\imath \sin(k_2 h) U_{i,M} - U_{i,M-1} = 0, \quad i = 0, \dots, N. \qquad (1.159)$$

Upon eliminating the fictitious nodal values $U_{N+1,j}$, $U_{i,M+1}$ and $U_{N+1,M+1}$ from Eqs. (1.95), (1.158), (1.159) we obtain

$$2U_{N-1,j} + (-4A(kh) + 2\imath \sin(k_1 h)) U_{N,j} + U_{N,j-1} + U_{N,j+1} = 0,$$

$$U_{i-1,M} + (-4A(kh) + 2\imath \sin(k_2 h)) U_{i,M} + U_{i+1,M} + 2U_{i,M-1} = 0, \qquad (1.160)$$

$$i = 1, \dots, N-1, \quad j = 1, \dots, M-1,$$

$$2U_{N-1,M} + (-4A(kh) + 2\imath \sin(k_1 h) + 2\imath \sin(k_2 h)) U_{N,M} + 2U_{N,M-1} = 0.$$

Now using a master index l

$$\hat{u}_l = U_{i,j}, \quad l = (j-1)N + i, \quad i = 1, \dots, N, \quad j = 1, \dots, M, \qquad (1.161)$$

the two-dimensional method (1.95) leads to the system of algebraic equations

$$\mathbf{AU} = \mathbf{F}. \qquad (1.162)$$

Here $\mathbf{A}$ is a nonsymmetric band matrix of dimension $(NM) \times (NM)$; lower and upper bandwidths equal to N, but only mean diagonal, first and Nth subdiagonal and superdiagonal

Table 1.11 The convergence of the two-dimensional method (1.95) with $T_0(kh)$ (1.96a), $J_0(kh)$ (1.96b) and $C(k_1h, k_2h)$ (1.96c) for the **Example** 1.7 versus the parameter k and the numbers of grid partitions N, M. The first column shows the values of the parameter k, the second ones displays the numbers of grid N, M. The third, fifth and seventh columns display the maximum absolute error $\|e_u\|_{\infty,h}$ for (1.96a), (1.96b) and (1.96c), respectively, while the fourth and sixth columns present corresponding Runge coefficients. The last three columns show the values of $T_0(kh)$, $J_0(kh)$ and $C(k_1h, k_2h)$. Here $\theta = \pi/4$

k	$N = M$	$T_0(kh)$ $\|e_u\|_{\infty,h}$	Runge	$J_0(kh)$ $\|e_u\|_{\infty,h}$	Runge	$C(k_1h, k_2h)$ $\|e_u\|_{\infty,h}$	$T_0(kh)$	$J_0(kh)$	$C(k_1h, k_2h)$
10	8	4.52(−01)		2.07(−01)		2.14(−33)	0.609	0.645	0.634
	16	1.02(−01)	4.40	5.01(−02)	4.14	5.55(−33)	0.902	0.904	0.903
	32	2.50(−02)	4.10	1.24(−02)	4.03	1.29(−32)	0.975	0.975	0.975
	64	6.21(−03)	4.02	3.10(−03)	4.00	2.25(−32)	0.993	0.993	0.993
	128	1.55(−03)	4.00	7.74(−04)	4.00	2.25(−32)	0.998	0.998	0.998
	256	3.87(−04)	4.00	1.93(−04)	4.00	2.25(−32)	0.999	0.999	0.999
30	8	1.00(−00)		4.17(−00)		5.10(−33)	−2.515	−0.401	−0.882
	16	2.12(−00)	0.47	1.41(−00)	2.94	7.22(−33)	0.121	0.296	0.242
	32	7.64(−01)	2.77	3.61(−01)	3.92	1.11(−32)	0.780	0.792	0.788
	64	1.82(−01)	4.19	8.93(−02)	4.04	3.08(−32)	0.945	0.945	0.945
	128	4.47(−02)	4.06	2.22(−02)	4.01	3.08(−32)	0.986	0.986	0.986
	256	1.11(−02)	4.01	5.55(−03)	4.00	3.08(−32)	0.996	0.996	0.996
50	8	1.00(−00)		1.91(−00)		5.57(−33)	−8.765	0.213	−0.288
	16	1.02(−00)	0.97	2.86(−00)	0.66	6.80(−33)	−1.441	−0.299	−0.596
	32	2.18(−00)	0.47	1.52(−00)	1.87	2.05(−32)	0.389	0.476	0.449
	64	8.74(−01)	2.49	4.21(−01)	3.62	2.78(−32)	0.847	0.853	0.851
	128	2.13(−01)	4.09	2.91(−01)	3.81	3.08(−32)	0.961	0.962	0.962
	256	5.27(−02)	4.04	2.62(−02)	4.00	3.08(−32)	0.990	0.990	0.990
70	8	1.02(−00)		4.39(−00)		1.39(−32)	−18.140	−0.025	0.995
	16	1.00(−00)	1.02	1.38(+01)	0.31	1.29(−32)	−3.785	−0.347	−0.998
	32	2.04(−00)	0.48	2.01(−00)	6.88	1.73(−32)	−0.196	0.117	0.023
	64	2.05(−00)	0.99	1.11(−00)	1.81	3.44(−32)	0.700	0.722	0.715
	128	5.96(−01)	3.44	2.91(−01)	3.81	3.08(−32)	0.925	0.926	0.926
	256	1.46(−01)	4.05	7.28(−02)	3.99	3.08(−32)	0.981	0.981	0.981

Table 1.12 The same as in Table 1.11, but $\theta = \pi/8$

k	$N = M$	$T_0(kh)$		$J_0(kh)$		$C(k_1h, k_2h)$	$T_0(kh)$	$J_0(kh)$	$C(k_1h, k_2h)$
		$\|e_u\|_{\infty,h}$	Runge	$\|e_u\|_{\infty,h}$	Runge	$\|e_u\|_{\infty,h}$			
10	8	6.77(−01)		2.07(−05)		1.38(−33)	0.609	0.645	0.645
	16	1.51(−01)	4.47	2.87(−07)	72.10	4.74(−33)	0.902	0.904	0.904
	32	3.68(−02)	4.10	4.36(−09)	65.85	9.10(−33)	0.975	0.975	0.975
	64	9.16(−03)	4.02	6.77(−11)	64.44	2.82(−32)	0.993	0.993	0.993
	128	2.28(−03)	4.00	1.05(−12)	64.10	7.83(−32)	0.998	0.998	0.998
	256	5.71(−04)	4.00	1.65(−14)	64.02	5.73(−32)	0.999	0.999	0.999
30	8	1.02(−00)		1.08(−01)		4.09(−33)	−2.515	−0.401	−0.406
	16	2.22(−00)	0.45	8.95(−04)	120.75	4.72(−33)	0.121	0.296	0.296
	32	1.00(−00)	2.21	9.86(−06)	90.79	1.09(−32)	0.780	0.792	0.792
	64	2.39(−01)	4.20	1.43(−07)	68.92	2.28(−32)	0.945	0.945	0.945
	128	5.87(−02)	4.06	2.19(−09)	65.10	7.29(−32)	0.986	0.986	0.986
	256	1.46(−02)	4.01	3.42(−11)	64.26	3.08(−32)	0.996	0.996	0.996
50	8	1.02(−00)		2.69(−00)		8.39(−33)	−8.765	0.213	0.070
	16	1.13(−00)	0.90	4.91(−02)	54.81	6.87(−33)	−1.441	−0.299	−0.300
	32	2.32(−00)	0.48	4.26(−04)	115.16	1.23(−32)	0.389	0.476	0.476
	64	1.10(−00)	2.09	5.29(−06)	80.62	3.46(−32)	0.847	0.853	0.853
	128	2.72(−01)	4.06	7.85(−08)	67.39	3.08(−32)	0.961	0.962	0.962
	256	6.72(−02)	4.04	1.21(−09)	64.78	3.08(−32)	0.990	0.990	0.990
70	8	1.00(−00)		7.27(−00)		8.29(−33)	−18.140	−0.025	−0.603
	16	1.00(−00)	0.99	8.82(−01)	8.25	1.93(−32)	−3.785	−0.347	−0.362
	32	2.09(−00)	0.48	6.39(−03)	138.06	1.56(−32)	−0.196	0.117	0.117
	64	2.26(−00)	0.92	5.96(−05)	107.08	2.34(−32)	0.700	0.722	0.722
	128	7.44(−01)	3.04	8.38(−07)	71.20	3.08(−32)	0.925	0.926	0.926
	256	1.83(−01)	4.03	1.27(−08)	65.59	3.08(−32)	0.981	0.981	0.981

elements are nonzero; the vector $\mathbf{F}$ contains the boundary conditions of Eq. (1.157), and its first N elements and $(jN+1)$th $(j=2,\ldots,M)$ elements are nonzero. For example, below presented Eq. (1.162) at $N=M=4$:

$$
\left(
\begin{array}{cccc|cccc|cccc|cccc}
D & 1 & & & 1 & & & & & & & & & & & \\
1 & D & 1 & & & 1 & & & & & & & & & & \\
 & 1 & D & 1 & & & 1 & & & & & & & & & \\
 & & 2 & S & & & & 1 & & & & & & & & \\
\hline
1 & & & & D & 1 & & & 1 & & & & & & & \\
 & 1 & & & 1 & D & 1 & & & 1 & & & & & & \\
 & & 1 & & & 1 & D & 1 & & & 1 & & & & & \\
 & & & 1 & & & 2 & S & & & & 1 & & & & \\
\hline
 & & & & 1 & & & & D & 1 & & & 1 & & & \\
 & & & & & 1 & & & 1 & D & 1 & & & 1 & & \\
 & & & & & & 1 & & & 1 & D & 1 & & & 1 & \\
 & & & & & & & 1 & & & 2 & S & & & & 1 \\
\hline
 & & & & & & & & 2 & & & & Q & 1 & & \\
 & & & & & & & & & 2 & & & 1 & Q & 1 & \\
 & & & & & & & & & & 2 & & & 1 & Q & 1 \\
 & & & & & & & & & & & 2 & & & 2 & Y
\end{array}
\right)
\left(
\begin{array}{c}
\hat{u}_1 \\ \hat{u}_2 \\ \hat{u}_3 \\ \hat{u}_4 \\ \hline \hat{u}_5 \\ \hat{u}_6 \\ \hat{u}_7 \\ \hat{u}_8 \\ \hline \hat{u}_9 \\ \hat{u}_{10} \\ \hat{u}_{11} \\ \hat{u}_{12} \\ \hline \hat{u}_{13} \\ \hat{u}_{14} \\ \hat{u}_{15} \\ \hat{u}_{16}
\end{array}
\right)
= -
\left(
\begin{array}{c}
f_2(h)+f_1(h) \\ f_2(2h) \\ f_2(3h) \\ f_2(4h) \\ \hline f_1(2h) \\ \\ \\ \\ \hline f_1(3h) \\ \\ \\ \\ \hline f_1(4h) \\ \\ \\
\end{array}
\right), \quad (1.163)
$$

where $D=-4A(kh)$, $S=D+2\imath\sin(k_1h)$, $Q=D+2\imath\sin(k_2h)$, $Y=D+2\imath\sin(k_2h)+2\imath\sin(k_2h)$.

The corresponding the maximum absolute error $\|e_u\|_{\infty,h}$ and Runge coefficients versus the number of grids N and M, and the value of k at $\theta=\pi/4$ are displayed in Table 1.11. When we use two-dimensional method (1.95) with $C(k_1h, k_2h)$, the maximum absolute errors $\|e_u\|_{\infty,h}$ are less then 10^{-30} for each number of grid $N=M$ and the parameter k (quadruple-precision arithmetic on Intel FORTRAN Compiler). The last three columns of the Table 1.11 given the comparison of $T_0(kh)$, $J_0(kh)$ and $C(k_1h, k_2h)$.

The last, we chosen $k_1=k\cos(\theta)$ with $\theta=\pi/8$. From Table 1.12, we observed the convergence of the method (1.95) with $J_0(kh)$ increased up to $O((kh)^6)$, that corresponds to the theoretical estimation (1.108). We obtained the similar results at $\theta=3\pi/8$, $5\pi/8$, $7\pi/8$.

References

1. R.P. Agarwal, *Difference Equations and Inequalities: Theory, Methods and Applications*, 2nd edn. (CRC Press, Boca Raton, 2000)
2. R.E. Mickens, *Nonstandard finite Difference Models of Differential Equations* (World Scientific, Singapore, 1994)
3. A.A. Samarskii, *Theory of Difference Equations* (Nauka, Moscow, 1977)

4. T. Zhanlav, V. Ulziibayar, The best finite difference scheme for the Helmholtz equation. Am. J. Comput. Math. **2**, 207–212 (2012)
5. V. Ulziibayar, High-order finite-difference schemes for numerical solution of some partial differential equations, Doctoral thesis, Ulaanbaatar, Mongolia (2014)
6. B. Batgerel, T. Zhanlav, An exact finite-difference scheme for Sturm-Louiville problems. Sc. Trans. NUM **1**(120), 8–15 (1996)
7. L. Bao, L.W. Wei, S. Zhao, Numerical solution of the Helmholtz equation with high wavenumbers. Int. J. Numer. Methods Eng. **59**, 389–408 (2004)
8. Y.S. Wong, L. Li, Exact finite difference schemes for solving Helmholtz equation at any wavenumber. Int. J. Numer. Anal. Model. Ser. B **2**, 91–108 (2011)
9. A.L. Larry, R. Luczak, J.W. Nehrbass, A new finite difference method for the Helmholtz equation using symbolic computation. Int. J. Comput. Eng. **4**, 121–144 (2003)
10. J.B.R. Do Val, M.L. Andrade Fo, The numerical solution of the Dirichlet problem for the Helmholtz equation. Appl. Math. Lett. **9**, 85–89 (1996)
11. M.L. Andrade, J.B.R. Do Val, A numerical scheme based on mean value solutions for the Helmholtz equation on triangular grids. Math. Comput. **66** 477–493 (1997)
12. I.S. Gradshteyn, I.M. Ryzhik, Table of integrals, series and products, 7Ed 2007, Academic Press is an imprint of Elsevier
13. I.V. Puzynin, T.L. Boyadjiev, S.I. Vinitsky, E.V. Zemlyanaya, T.P. Puzynina, O. Chuluunbaatar, Methods of computational physics for investigation of models of complex physical systems. Phys. Part. Nucl. **38**, 70–116 (2007)
14. T. Zhanlav, R. Mijiddorj, O. Chuluunbaatar, The continuous analogue of Newton's method for solving eigenvalues and eigenvectors of matrices. Bull. Tver State Univ. **14**, 27–37 (2008). (in Russian)
15. D.Y. Kwak, J.S. Lee, The V-cycle multigrid convergence of some finite difference scheme for the Helmholtz equation. Sib. J. Numer. Math. **8**, 207–218 (2005)

Higher-Order Finite-Difference Methods for the Burgers' Equations

Abstract

We propose several higher-order explicit finite difference methods (FDMs) for solving one- and two-dimensional Burgers' equations, as well as two-dimensional coupled Burgers' equations with a corresponding initial condition and boundary conditions. The proposed FDMs for solving one- and two-dimensional Burgers' equations have a sixth-order approximation in space variables and a third-order approximation in the time variable, while for solving two-dimensional coupled Burgers' equations, they have a fourth-order approximation in space variables and a second-order approximation in the time variable. A numerical method for solving one-dimensional Burgers' equation using the relationship between the heat equation and Burgers' equation is also presented. This method has a sixth-order approximation in the space variable. The accuracy of the proposed method is demonstrated by some test problems. The numerical results are in good agreement with the exact solutions.

2.1 Introduction

Burgers' equation is an important non-linear parabolic partial differential equation widely used to model various physical processes such as fluid flows [1–3], turbulence, boundary layer behavior, shock wave formation and mass transport [4]. Consequently, it is one of the principal model equations used to test the accuracy of new numerical methods. There are various numerical methods for solving Burgers' equation with corresponding initial condition and boundary conditions [5–10]. In Ref. [5], an efficient numerical method based on Haar wavelets and the quasi linearization process is developed for solving nonlinear one-dimensional Burgers' equation with Dirichlet boundary conditions. The distributed approximating functional method is applied for solving one-dimensional Burgers' equation and

© The Author(s), under exclusive license to Springer Nature Switzerland AG 2024   35
U. Vandandoo et al., *High-Order Finite Difference and Finite Element Methods for Solving Some Partial Differential Equations*, Synthesis Lectures on Engineering, Science, and Technology, https://doi.org/10.1007/978-3-031-44784-6_2

unsteady Burgers' equation [6]. In [7], unsteady Burgers' equation with Dirichlet boundary conditions is solved by the lattice Boltzmann method. The high-order two-point compact alternating direction implicit algorithm is introduced to solve unsteady Burgers' equation. A comparison of the proposed method with the fourth-order DuFort-Frankel [8] method is constructed in terms of accuracy and computational efficiency. In Ref. [9], modified Bi-quintic B-spline basis functions are proposed and applied to unsteady Burgers' equation using the Galerkin method to obtain its numerical solution. However, in some cases, the high-order truncation error of typical methods can become very large and therefore cannot be neglected. This implies that attempting to use high-order methods for solving equations that exhibit erratic, turbulent-like solutions may fail to provide the expected results. This must be kept in mind.

Several researchers have been interested in studying the properties of twodimensional coupled Burgers' equation (TDCBE) using various numerical techniques. There exist many different explicit and implicit numerical methods with second-order approximation in space variables, and a first- or second-order approximation in the time variable. For example, in [3, 11–13], the Crank-Nicolson method using different fully implicit or semi-implicit FDMs for the numerical solution of the TDCBE is applied. Implicit logarithmic and local discontinuous Galerkin FDMs for the numerical solution of the TDCBE are proposed in [14, 15]. In addition, in [3], an explicit method using the FDM is applied. Implicit FDMs with fourth-order approximation in space variables, and second-order approximation in time variable are proposed in [16, 17]. These methods are based on the Crank-Nicolson method with the PadÂ´e approximation of the finite difference operator and the hybrid Crank-Nicolson-Du Fort and Frankel method, respectively. However, implicit methods on each time layer require the solution of an algebraic system. In the multidimensional case of heat equations, a large calculation time is required to solve the algebraic systems up to the final time layer, even taking into account the band structure of the matrices [18].

2.2   High-Order Numerical Solution of the One-Dimensional Burgers' Equation

The Burgers' equation can be considered as an approach to the Navier-Stokes equations [19, 20]. Since both contain nonlinear terms of the type: unknown functions multiplied by a first derivative and both contain high-order terms multiplied by a small parameter. On the other hand, the Burgers' equation is one of a few nonlinear equations which can be solved exactly for an arbitrary initial and boundary conditions [2]. However these exact solutions are impractical for the small values of viscosity constant due to a slow convergence of series solutions. Thus many numerical methods are constructed for a numerical solution of the Burgers' equation for small values of viscosity constant which corresponds to a steep front in the propagation of dynamic wave forms [2, 21–25]. The study of the general properties of the Burgers' equation has motivated considerable attention due to its applications in field as diverse as number theory, gas dynamics, heat conduction, elasticity, etc. [2].

We consider a one-dimensional quasi-linear parabolic partial differential equation [10]

$$\frac{\partial u}{\partial t} + u \frac{\partial u}{\partial x} = v \frac{\partial^2 u}{\partial x^2}, \quad a < x < b, \quad t > 0, \tag{2.1}$$

with an initial condition

$$u(x, 0) = \varphi(x), \quad a < x < b, \tag{2.2}$$

and Dirichlet boundary conditions

$$u(a, t) = u(b, t) = 0, \quad t > 0, \tag{2.3}$$

where $v > 0$ is a coefficient of the kinematic viscosity and $\varphi(x)$ is known function.

It is well known that, by the Hopf-Cole transformation

$$u(x, t) = -2v \frac{\theta'_x(x, t)}{\theta(x, t)}, \tag{2.4}$$

the Burgers' equation transforms to the linear heat equation

$$\frac{\partial \theta(x, t)}{\partial t} = v \frac{\partial^2 \theta(x, t)}{\partial x^2}, \quad a < x < b, \quad t > 0, \tag{2.5}$$

with initial condition

$$\theta(x, 0) = \exp\left(-\frac{1}{2v} \int_a^x \varphi(\xi) d\xi\right), \quad a < x < b, \tag{2.6}$$

and Neumann boundary conditions

$$\theta'_x(a, t) = \theta'_x(b, t) = 0, \quad t > 0. \tag{2.7}$$

Symbol "$'$" denotes the derivative with respect to variable x. Thus, if $\theta(x, t)$ is any solution of the heat equation (2.5) subject to the conditions (2.6) and (2.7), then the function (2.4) is a solution of the Burgers' equation (2.1) with the conditions (2.2) and (2.3).

We assume that the numerical solution of the heat problem defined by Eqs. (2.5)–(2.7) is found by any of known methods with higher accuracy. For example, this problem can be solved by well-known Crank–Nicolson method [26] and a high-order explicit method proposed by Zhanlav in [27]

$$\theta_i^{n+1} = \frac{\beta - \gamma}{\beta + \gamma} \theta_i^{n-1} + \frac{\beta \gamma}{\beta + \gamma} \left(\theta_{i-1}^n - 2\theta_i^n + \theta_{i+1}^n\right) + \frac{2\gamma}{\beta + \gamma} \theta_i^n, \tag{2.8}$$

$$\gamma = \frac{2\tau v}{h^2}, \quad i = 1, \ldots, N - 1, \quad h = \frac{b - a}{N}, \quad n = 1, 2, \ldots.$$

Here θ_i^n is the approximate solution at the mesh points $(x_i = ih, t_n = n\tau)$, where h is a spatial step, τ is a time step. Using the Taylor expansions of $\theta_i^{n\pm1}$, $\theta_{i\pm1}^n$ at the point (x_i, t_n), and an identity

$$\frac{\partial^m \theta(x, t)}{\partial t^m} = v^m \frac{\partial^{2m}}{\partial x^{2m}} \theta(x, t), \quad m \geq 0, \tag{2.9}$$

we have the truncation error of the method (2.8)

$$\psi_i^n = \frac{\beta}{\beta + \gamma} \left(2v - \frac{\gamma}{\tau} h^2\right) \frac{\partial^2 \theta(x, t)}{\partial x^2} + \frac{\gamma}{\tau} \frac{1}{\beta + \gamma} \left(\tau^2 v^2 - \beta \frac{h^4}{12}\right) \frac{\partial^4 \theta(x, t)}{\partial x^4}$$
$$+ \frac{\beta}{\beta + \gamma} \left(\frac{\tau^2 v^3}{3} - \frac{\gamma}{\tau} \frac{h^6}{360}\right) \frac{\partial^6 \theta(x, t)}{\partial x^6} + O(\tau^3 + h^6). \tag{2.10}$$

Equating the coefficients of the partial derivatives to zero in (2.10), we obtain following system of equations

$$\begin{cases} 2\tau v - \gamma h^2 = 0, \\ 12\tau^2 v^2 - \beta h^4 \equiv 3\gamma^2 h^2 - \beta h^2 = 0, \\ 120\tau^3 v^3 - \gamma h^6 \equiv 15\gamma^3 h^6 - \gamma h^6 = 0. \end{cases} \tag{2.11}$$

From here we obtained

$$\gamma = \frac{2\tau v}{h^2} = \frac{1}{\sqrt{15}}, \quad \beta = 3\gamma^2 = \frac{1}{5}. \tag{2.12}$$

It means that truncation error of the method (2.8) with conditions (2.12) is of the order $O(\tau^3 + h^6)$. When $\beta = 1$, the method (2.8) leads to the well-known Du Fort–Frankel's one [26].

For finding the stability condition of the method (2.8), we seek the partial solution in the form:

$$\theta_{i,j}^n = q^n \exp(\iota i h \psi). \tag{2.13}$$

From (2.8) we have

$$(\beta + \gamma)q^2 + 2bq + \gamma - \beta = 0, \tag{2.14}$$

$$b = -\beta\gamma(\cos(h\psi) - 1) - \gamma. \tag{2.15}$$

We have following theorem:

Theorem 2.1 *[18] Let $a > 0$, b and c are real numbers. Then roots of a quadratic equation $aq^2 + 2bq + c = 0$ satisfy to the condition $|q_{1,2}| \leq 1$ if and only if*

$$\frac{|c|}{a} \leq 1, \quad 2|b| \leq a + c. \tag{2.16}$$

Using the conditions (2.16), we obtain

$$
\begin{cases}
\frac{|c|}{a} = \frac{|\gamma - \beta|}{\gamma + \beta} \leq 1, \\
a + c = 2\gamma \geq -2b = 2\beta\gamma(\cos(h\psi) - 1) + 2\gamma, \\
a + c = 2\gamma \geq 2b = -2\beta\gamma(\cos(h\psi) - 1) - 2\gamma,
\end{cases}
\Rightarrow
\begin{cases}
0 \leq \gamma, \; 0 \leq \beta, \\
1 \geq \cos(h\psi), \\
\frac{2}{\beta} - 1 \geq -\cos(h\psi).
\end{cases}
\tag{2.17}
$$

The last inequality is true for $\beta \leq 1$.

It should be mentioned that the method (2.8) is a three-level one in time. Hence, in order to find θ_i^n at level two it requires two values θ_i^n at level 0 and 1, i.e., θ_i^0 and θ_i^1. Using the Taylor expansion of $\theta(x, \tau)$ at point $(x, 0)$ and Eq. (2.5), we obtain

$$
\theta(x, \tau) = \theta(x, 0) + \nu \frac{\partial^2 \theta(x, 0)}{\partial^2 x} \tau + \frac{\nu^2}{2} \frac{\partial^4 \theta(x, 0)}{\partial^4 x} \tau^2 + O(\tau^3).
\tag{2.18}
$$

From Eq. (2.18) we will find θ_i^1.

2.2.1   Construction of High-Order Finite-Difference Methods

We suppose that the solution of Eqs. (2.5)–(2.7) is a sufficiently smooth function with respect to x and t. So, from the Taylor expansions of $\theta(x_{i+1}, t)$ and $\theta(x_{i-1}, t)$ at point (x_i, t) we have

$$
\frac{\theta(x_{i+1}, t) - \theta(x_{i-1}, t)}{2h} = \theta_x'(x_i, t) + \frac{\theta_x'''(x_i, t)}{6} h^2 + \frac{\theta_x^{(5)}(x_i, t)}{120} h^4 + O(h^6), \quad (2.19)
$$

$$
\frac{\theta_x'(x_{i+1}, t) - 2\theta_x'(x_i, t) + \theta_x'(x_{i-1}, t)}{h^2} = \theta_x'''(x_i, t) + \frac{\theta_x^{(5)}(x_i, t)}{12} h^2 + O(h^4). \quad (2.20)
$$

Eliminating θ_x''' from (2.19) and (2.20), we obtain

$$
\frac{\theta(x_{i+1}, t) - \theta(x_{i-1}, t)}{2h} = \frac{\theta_x'(x_{i+1}, t) + 4\theta_x'(x_i, t) + \theta_x'(x_{i-1}, t)}{6}
$$
$$
- \frac{\theta_x^{(5)}(x_i, t)}{180} h^4 + O(h^6).
\tag{2.21}
$$

Omitting the small term in the right-hand side of the obtained FDM:

$$
\frac{\theta_{i+1}^n - \theta_{i-1}^n}{2h} = \frac{(\theta_x')_{i+1}^n + 4(\theta_x')_i^n + (\theta_x')_{i-1}^n}{6}, \quad i = 1, 2, \ldots, N - 1.
\tag{2.22}
$$

The truncation error of this method is $O(h^4)$. Finding θ_x' from (2.4) and substituting it into (2.22), we obtain a compact FDM for approximate solution $y_i^n \equiv y(x_i, t_n)$ of $u(x_i, t_n)$:

$$\theta_{i-1}^n y_{i-1}^n + 4\theta_i^n y_i^n + \theta_{i+1}^n y_{i+1}^n = -\frac{6v}{h}\left(\theta_{i+1}^n - \theta_{i-1}^n\right),\tag{2.23}$$
$$i = 1, 2, \ldots, N-1, \quad n = 1, 2, \ldots,$$

with boundary conditions

$$y_0^n = y_N^n = 0.\tag{2.24}$$

If we denote $\theta_i^n y_i^n$ by v_i^n, then the method (2.23), (2.24) leads to

$$v_{i-1}^n + 4v_i^n + v_{i+1}^n = -\frac{6v}{h}\left(\theta_{i+1}^n - \theta_{i-1}^n\right),\tag{2.25}$$
$$v_0^n = v_N^n = 0.\tag{2.26}$$

The last system has a unique solution set $\left(v_0^n, v_1^n, \ldots, v_N^n\right)$ since its the matrix is diagonally dominant. It means that the tridiagonal system (2.23), (2.24) has a unique solution set $(y_0^n, y_1^n, \ldots, y_N^n)$ for each $n = 1, 2, \ldots$, and it can be solved by efficient elimination method [28]. Moreover, it is also possible to obtain a high-order FDM than (2.23), (2.24).

Using the Taylor expansions of $\theta(x_{i+2}, t)$ and $\theta(x_{i-2}, t)$ at the point (x_i, t) we have

$$\frac{\theta(x_{i+2}, t) - \theta(x_{i-2}, t)}{4h} = \theta_x'(x_i, t) + \frac{\theta_x'''(x_i, t)}{6}4h^2 + \frac{\theta_x^{(5)}(x_i, t)}{120}16h^4 + O(h^6).\tag{2.27}$$

We can eliminate the term with $\theta^{(5)}(x_i, t)$ from (2.27) and (2.19). As a result we have

$$16\frac{\theta(x_{i+1}, t) - \theta(x_{i-1}, t)}{2h} - \frac{\theta(x_{i+2}, t) - \theta(x_{i-2}, t)}{4h}$$
$$= 15\theta_x'(x_i, t) + 2\theta_x'''(x_i, t)h^2 + O(h^6).\tag{2.28}$$

We also use the well-known five-point approximation formula for $\theta_x'''(x_i, t)$

$$\theta_{x,i}''' = \frac{1}{12h^2}\left(-\theta_{x,i-2}' + 16\theta_{x,i-1}' - 30\theta_{x,i}' + 16\theta_{x,i+1}' - \theta_{x,i+2}'\right) + O(h^4),\tag{2.29}$$

which holds for sufficiently smooth function $\theta(x, t)$ with respect to x variable. Substituting (2.29) into (2.28) and using the Hopf-Cole transformation given by Eq. (2.4) we obtain

$$v_{i-2}^n - 16v_{i-1}^n - 60v_i^n - 16v_{i+1}^n + v_{i+2}^n = c_i^n,\tag{2.30}$$
$$c_i^n = -\frac{3v}{h}\left(-\theta_{i-2}^n + 32\theta_{i-1}^n - 32\theta_{i+1}^n + \theta_{i+2}^n\right), \quad i = 2, 3, \ldots, N-2.$$

Of course, besides of $v_0^n = v_N^n = 0$ we need additionally two end conditions v_1^n and v_{N-1}^n in order to solve the system (2.30). We differentiate Eq. (2.5) $(2k-1)$-times with respect to x and find that

$$\theta_x^{(2k+1)}(x, t) = \frac{1}{v}\frac{\partial}{\partial t}\theta_x^{(2k-1)}(x, t), \quad k = 1, 2, \ldots.\tag{2.31}$$

From the Neumann boundary conditions (2.7) it is obvious that

$$\theta_x^{(2k+1)}(x_0, t) = \theta_x^{(2k+1)}(x_N, t) = 0, \quad k = 0, 1, \dots. \tag{2.32}$$

Then from (2.19) it follows that

$$\theta(x_1, t) = \theta(x_{-1}, t), \quad x_1 = h, \quad x_{-1} = -h. \tag{2.33}$$

Also we differentiate Eq. (2.4) $(2k)$-times with respect to x and find that

$$\theta_x^{(2k+1)} = -\frac{1}{2v} v_x^{(2k)}(x, t), \quad k = 0, 1, \dots. \tag{2.34}$$

If we use (2.32) in (2.34), then we obtain

$$v_x^{(2k)}(x_0, t) = v_x^{(2k)}(x_N, t) = 0. \tag{2.35}$$

From the Taylor expansions

$$v(x_{-1}, t) = v(x_0, t) - v_x'(x_0, t)h + \frac{v_x''(x_0, t)}{2}h^2 - \frac{v_x'''(x_0, t)}{6}h^3 + \cdots,$$

$$v(x_1, t) = v(x_0, t) + v_x'(x_0, t)h + \frac{v_x''(x_0, t)}{2}h^2 + \frac{v_x'''(x_0, t)}{6}h^3 + \cdots, \tag{2.36}$$

$$v(x_{N-1}, t) = v(x_N, t) - v_x'(x_N, t)h + \frac{v_x''(x_N, t)}{2}h^2 - \frac{v_x'''(x_N, t)}{6}h^3 + \cdots,$$

$$v(x_{N+1}, t) = v(x_N, t) + v_x'(x_N, t)h + \frac{v_x''(x_N, t)}{2}h^2 + \frac{v_x'''(x_N, t)}{6}h^3 + \cdots,$$

and (2.35) we conclude that

$$v(x_{-1}, t) = -v(x_1, t), \quad v(x_{N-1}, t) = -v(x_{N+1}, t). \tag{2.37}$$

Hence, taking into account (2.33) and (2.37), the FDM (2.30) has the forms for $i = 1, N - 1$

$$-61v_1^n - 16v_2^n + v_3^n = -\frac{3v}{h} \left(32\theta_0^n - \theta_1^n - 32\theta_2^n + \theta_3^n \right), \tag{2.38}$$

$$v_{N-3}^n - 16v_{N-2}^n - 61v_{N-1}^n = -\frac{3v}{h} \left(-\theta_{N-3}^n + 32\theta_{N-2}^n + \theta_{N-1}^n - 32\theta_N^n \right).$$

Thus, we have FDM (2.30), (2.38) with truncation error $O(h^6)$.

The solution procedure of system (2.30), (2.38) is essentially simplified by using Z-folding algorithm [29]. Namely, if we use notation

$$Z_i^n = v_{i-1}^n + av_i^n + v_{i+1}^n, \tag{2.39}$$

then it is easy to show that the Eq. (2.30) can be re written as

$$Z_{i-1}^n + bZ_i^n + Z_{i+1}^n = c_i^n, \quad i = 2, \ldots, N-2, \tag{2.40}$$

under conditions

$$b = -8 \pm 3\sqrt{14}, \quad a = -8 \mp 3\sqrt{14}. \tag{2.41}$$

It means, that the solution of penta-diagonal system (2.30) leads to two three-diagonal systems (2.40) and (2.39) consequently, both of which has a diagonally dominance [30].

Now, we consider z_0^n and z_N^n given by (2.39)

$$Z_0^n = v(x_{-1}, t_n) + av(x_0, t_n) + v(x_1, t_n), \tag{2.42}$$

$$Z_N^n = v(x_{N-1}, t_n) + av(x_N, t_n) + v(x_{N+1}, t_n). \tag{2.43}$$

By (2.37) and (2.26) we have

$$Z_0^n \equiv z(x_0, t_n) = av(x_0, t_n) = 0, \quad Z_N^n \equiv z(x_N, t_n) = av(x_N, t_n) = 0. \tag{2.44}$$

Thus, we obtain the system

$$Z_{i-1}^n + bZ_i^n + Z_{i+1}^n = c_i^n, \quad i = 1, \ldots, N-1, \tag{2.45}$$

$$Z_0^n = Z_N^n = 0.$$

After solving the last system, one can solve (2.39), i.e.

$$v_{i-1}^n + av_i^n + v_{i+1}^n = Z_i^n, \quad i = 1, \ldots, N-1, \tag{2.46}$$

$$v_0^n = v_N^n = 0,$$

and thereby we obtain

$$u_i^n = \frac{v_i^n}{\theta_i^n}, \quad i = 0, \ldots, N. \tag{2.47}$$

2.2.2   Numerical Results

In this section we demonstrate the accuracy of the proposed FDMs (2.8), (2.12), (2.45), (2.46) by solving exact solvable problems and compare the numerical results with the existing results. The computations are performed using the MatLab program.

Example 2.1 We consider the Burgers' equation (2.1) with the initial condition

$$u(x, 0) = \sin(\pi x), \quad 0 < x < 1, \tag{2.48}$$

and the Dirichlet boundary conditions

$$u(0, t) = u(1, t) = 0, \quad t > 0. \tag{2.49}$$

The corresponding heat equation obeys the following initial condition

$$\theta(x, 0) = \exp\left(-\frac{1 - \cos(\pi x)}{2\pi v}\right), \tag{2.50}$$

and the Neumann boundary conditions

$$\theta'_x(0, t) = \theta'_x(1, t) = 0, \quad t > 0. \tag{2.51}$$

The Fourier series solution to the above heat problem defined by Eqs. (2.5)–(2.7) can obtained easily as [2]

$$\theta(x, t) = a_0 + \sum_{n=1}^{\infty} a_n \exp(-n^2\pi^2 vt) \cos(n\pi x), \tag{2.52}$$

with the Fourier coefficients

$$a_0 = \int_0^1 \theta(x, 0)dx, \quad a_n = 2\int_0^1 \theta(x, 0)\cos(n\pi x)dx, \quad n = 1, 2, 3, \ldots. \tag{2.53}$$

Therefore, the Fourier solution to the problem given by Eqs. (2.1), (2.48) and (2.49) is obtained as [2]

$$u(x, t) = 2\pi v \frac{\sum_{n=1}^{\infty} a_n \exp(-n^2\pi^2 vt)n \sin(n\pi x)}{a_0 + \sum_{n=1}^{\infty} a_n \exp(-n^2\pi^2 vt)\cos(n\pi x)}. \tag{2.54}$$

Thus, we obtain the numerical solution of Burgers' equation (2.1), (2.48), (2.49) with higher accuracy provided that the solution θ_i^n of heat equation (2.5)–(2.7) is found with higher accuracy. Using (2.52) and (2.54) we find the truncation errors $\psi_1(x_i, t)$, $\psi_2(x_i, t)$ of methods (2.23) and (2.30), respectively, as follows

$$\psi_1(x_i, t) = v(x_{i-1}, t) + 4v(x_i, t) + v(x_{i+1}, t) + \frac{6v}{h}(\theta(x_{i+1}, t) - \theta(x_{i-1}, t))$$

$$= 4\pi v \sum_{n=1}^{\infty} a_n \exp(-n^2\pi^2 vt)n \sin(\pi n x_i)B_{n,1}(\alpha), \tag{2.55}$$

$$\psi_2(x_i, t) = v(x_{i-2}, t) - 16v(x_{i-1}, t) - 60v(x_i, t) - 16v(x_{i+1}, t) + v(x_{i+2}, t)$$

$$+\frac{3v}{h}\left[32(\theta(x_{i-1}, t) - \theta(x_{i+1}, t)) + \theta(x_{i+2}, t) - \theta(x_{i-2}, t)\right]$$

$$= 4\pi v \sum_{n=1}^{\infty} a_n \exp(-n^2\pi^2 vt)n \sin(\pi n x_i)B_{n,2}(\alpha), \tag{2.56}$$

Table 2.1 Convergence of the proposed methods for the numerical solution $y(x_i, T)$ to the exact solution $u(x_i, T)$ of **Example** 2.1 versus the number of nodes N. Here $\nu = 1$ and $T = (10\sqrt{15})^{-1}$

x	Numerical solution				Exact solution
	$N = 10$	$N = 20$	$N = 40$	$N = 80$	
0.1	0.228649	0.22865030	0.2286503154	0.2286503156451	0.2286503156477
0.2	0.437766	0.43776771	0.4377677345	0.4377677347901	0.4377677347942
0.3	0.608777	0.60877832	0.6087783451	0.6087783454149	0.6087783454190
0.4	0.725196	0.72519674	0.7251967567	0.7251967569572	0.7251967569600
0.5	0.774045	0.77404614	0.7740461512	0.7740461512588	0.7740461512595
0.6	0.747568	0.74756837	0.7475683734	0.7475683733302	0.7475683733289
0.7	0.645016	0.64501619	0.6450161825	0.6450161823893	0.6450161823870
0.8	0.474055	0.47405491	0.4740549069	0.4740549067930	0.4740549067907
0.9	0.251102	0.25110176	0.2511017581	0.2511017580559	0.2511017580546

where

$$B_{n,1}(\alpha) = 2 + \cos(\alpha) - 3\frac{\sin(\alpha)}{\alpha}, \quad \alpha = \pi n h,$$

$$B_{n,2}(\alpha) = 2\cos^2(\alpha) - 16\cos(\alpha) - 31 + 3(16 - \cos(\alpha))\frac{\sin(\alpha)}{\alpha}. \quad (2.57)$$

By using the Taylor expansions of functions $\cos\alpha$ and $\sin(\alpha)$ it is easy to show that

$$B_{n,1}(\alpha) = \frac{\alpha^4}{60}\left(1 - \frac{\alpha^2}{21} + O(\alpha^4)\right) = O(\alpha^4),$$

$$B_{n,2}(\alpha) = -\frac{4}{105}\alpha^6 + \frac{1}{252}\alpha^8 + O(\alpha^{10}) = O(\alpha^6). \quad (2.58)$$

Tables 2.1, 2.3 and 2.5 display convergence of the proposed methods for the numerical solution $y(x_i, T)$ to the exact solution $u(x_i, T)$ versus the number of nodes N for $\nu = 1$, $\nu = 0.1$ and $\nu = 0.01$ with $T = \frac{1}{10\sqrt{15}}$, $T = \frac{1}{\sqrt{15}}$ and $T = 10\sqrt{15}$. Table 2.7 presents the numerical results obtained by using the fourth-order weighted method [29, 30] and the exact solution for $\nu = 1$ at $T = \frac{1}{10\sqrt{15}}$. It is clearly observed that both numerical results are reasonably in good agreement with the exact solution. It is seen that for small values of ν, one must consider a large of N to obtain proper solution. The maximum absolute error $\|e\|_\infty = \max_{1\leq i\leq N-1} |y(x_i, T) - u(x_i, T)|$ versus the number of nodes N is displayed in Tables 2.2, 2.4, 2.6 and 2.8. It gives an approximate rate of convergence of the proposed methods and the weighted method. The errors are consistent with the theoretical expectations of $O(h^6)$ and $O(h^4)$.

Table 2.2 The maximum absolute error $\|e\|_\infty = \max\limits_{1 \le i \le N-1} |y(x_i, T) - u(x_i, T)|$ between numerical and exact solutions versus the number of nodes N, and corresponding Runge coefficients. The factor x in the brackets denotes 10^x

N	$\|e\|_\infty$	$\|e\|_\infty h / \|e\|_\infty h/2$
10	1.058410630083717(−06)	
20	1.679794564557469(−08)	63.008
40	2.635179296994750(−10)	63.744
80	4.136968545509490(−12)	63.698

Table 2.3 The same as in Table 2.1, but for $v = 0.1$ and $T = (\sqrt{15})^{-1}$

x	Numerical solution				Exact solution
	$N = 10$	$N = 20$	$N = 40$	$N = 80$	
0.1	0.157856	0.15786370	0.1578638152	0.1578638169663	0.1578638169929
0.2	0.311602	0.31161146	0.3116115963	0.3116115984406	0.3116115984745
0.3	0.456498	0.45650189	0.4565019578	0.4565019588173	0.4565019588320
0.4	0.586299	0.58629360	0.5862935224	0.5862935211388	0.5862935211184
0.5	0.691734	0.69172222	0.6917220370	0.6917220340560	0.6917220340080
0.6	0.757624	0.75761273	0.7576125552	0.7576125523766	0.7576125523305
0.7	0.757834	0.75783125	0.7578312008	0.7578312000723	0.7578312000594
0.8	0.650499	0.65050606	0.6505061712	0.6505061729091	0.6505061729349
0.9	0.391645	0.39165314	0.3916532762	0.3916532782669	0.3916532782993

Table 2.4 The same as in Table 2.2, but for Table 2.3

N	$\|e\|_\infty$	$\|e\|_\infty h / \|e\|_\infty h/2$
10	1.239220296311849(−05)	
20	1.957252674378296(−07)	63.314
40	3.066594644884901(−09)	63.824
80	4.795597252638117(−11)	63.946

Example 2.2 Consider the Burgers' equation (2.1) with the Dirichlet boundary conditions (2.49) and the initial condition

$$u(x, 0) = 4x(1 - x), \quad 0 < x < 1. \tag{2.59}$$

The problem (2.1), (2.49) and (2.59) has an exact solution and it is expressed by the formula (2.54). All needed coefficients a_n are calculated by (2.53) with

Table 2.5 The same as in Table 2.1, but $v = 0.01$ and $T = 10\sqrt{15}$

x	Numerical solution				Exact solution
	$N = 10$	$N = 20$	$N = 40$	$N = 80$	
0.1	0.00079050	0.00079045443	0.0007904544465	0.000790454446631	0.000790454446633
0.2	0.00151237	0.00151227042	0.0015122704379	0.001512270438210	0.001512270438214
0.3	0.00210061	0.00210047219	0.0021004722140	0.002100472214422	0.002100472214428
0.4	0.00249817	0.00249800187	0.0024980018957	0.002498001896186	0.002498001896192
0.5	0.00266108	0.00266089653	0.0026608965593	0.002660896559804	0.002660896559811
0.6	0.00256436	0.00256418989	0.0025641899229	0.002564189923350	0.002564189923357
0.7	0.00220777	0.00220761467	0.0022076146962	0.002207614696574	0.002207614696580
0.8	0.00161958	0.00161947230	0.0016194723165	0.001619472316759	0.001619472316762
0.9	0.00085680	0.00085673857	0.0008567385790	0.000856738579226	0.000856738579229

Table 2.6 The same as in Table 2.2, but for Table 2.5

N	$\|e\|_\infty$	$\|e\|_\infty h/\|e\|_\infty h/2$
10	1.840226970003904(−07)	
20	2.751935918726689(−11)	66.87
40	4.300379496946505(−13)	63.99

Table 2.7 The same as in Table 2.1, but for the fourth-order weighted method [29, 30] with $\sigma = 1/2 - h^2/(12v\tau)$

x	Numerical solution				Exact solution
	$N = 10$	$N = 20$	$N = 40$	$N = 80$	
0.1	0.228658	0.22865086	0.2286503500	0.2286503178	0.2286503156
0.2	0.437781	0.43776863	0.4377677926	0.4377677383	0.4377677347
0.3	0.608792	0.60877929	0.6087784054	0.6087783492	0.6087783454
0.4	0.725207	0.72519746	0.7251968015	0.7251967597	0.7251967569
0.5	0.774050	0.77404642	0.7740461683	0.7740461523	0.7740461512
0.6	0.747565	0.74756821	0.7475683638	0.7475683726	0.7475683733
0.7	0.645009	0.64501574	0.6450161541	0.6450161806	0.6450161823
0.8	0.474047	0.47405441	0.4740548752	0.4740549048	0.4740549067
0.9	0.251097	0.25110144	0.2511017381	0.2511017568	0.2511017580

$$\theta(x, 0) = \exp\left(\frac{2x^3 - 3x^2}{3v}\right), \quad 0 < x < 1. \tag{2.60}$$

The high-order numerical and exact solutions for $v = 1$, $v = 0.1$ and $v = 0.001$ at $T = 1/\sqrt{15}$ and $T = 10\sqrt{15}$ are presented Tables 2.9, 2.10 and 2.11.

Table 2.8 The same as in Table 2.2, but for Table 2.7

N	$\|e\|_\infty$	$\|e\|_\infty h / \|e\|_\infty h/2$
10	1.404989751185859(−05)	
20	9.509678503549779(−07)	14.774
40	6.056235823947986(−08)	15.702
80	3.802648973483258(−09)	15.926

Table 2.9 The same as in Table 2.1, but for **Example** 2.2 with $\nu = 1$ and $T = (\sqrt{15})^{-1}$

x	Numerical solution				Exact solution
	$N = 10$	$N = 20$	$N = 40$	$N = 80$	
0.1	0.02455883	0.02456346	0.02456374	0.02456376	0.02456376
0.2	0.04679722	0.04680605	0.04680659	0.04680662	0.04680663
0.3	0.06459077	0.06460299	0.06460374	0.06460379	0.06460379
0.4	0.07619922	0.07621368	0.07621457	0.07621462	0.07621463
0.5	0.08043556	0.08045088	0.08045183	0.08045188	0.08045189
0.6	0.07680068	0.07681537	0.07681627	0.07681633	0.07681633
0.7	0.06556400	0.06557658	0.06557735	0.06557740	0.06557740
0.8	0.04777049	0.04777969	0.04778025	0.04778029	0.04778029
0.9	0.02516037	0.02516522	0.02516552	0.02516554	0.02516554

Table 2.10 The same as in Table 2.9, but for $\nu = 0.1$ and $T = (\sqrt{15})^{-1}$

x	Numerical solution				Exact solution
	$N = 10$	$N = 20$	$N = 40$	$N = 80$	
0.1	0.16403609	0.16413204	0.16413714	0.16413745	0.16413747
0.2	0.32309362	0.32324755	0.32325614	0.32325666	0.32325669
0.3	0.47179502	0.47195307	0.47196269	0.47196329	0.47196333
0.4	0.60372599	0.60384776	0.60385627	0.60385682	0.60385685
0.5	0.71012965	0.71020720	0.71021370	0.71021413	0.71021416
0.6	0.77704033	0.77709963	0.77710480	0.77710514	0.77710517
0.7	0.77948836	0.77957486	0.77958062	0.77958098	0.77958101
0.8	0.67413875	0.67427966	0.67428743	0.67428790	0.67428793
0.9	0.40982075	0.40995677	0.40996384	0.40996427	0.40996429

Table 2.11 The same as in Table 2.9, but for $\nu = 0.0001$ and $T = 10\sqrt{15}$

x	Numerical solution				Exact solution
	$N = 10$	$N = 20$	$N = 40$	$N = 80$	
0.1	0.00269461	0.00257478	0.00256479	0.00256480	0.00256480
0.2	0.00538868	0.00514955	0.00512959	0.00512961	0.00512961
0.3	0.00808135	0.00772426	0.00769438	0.00769441	0.00769441
0.4	0.01077132	0.01029890	0.01025917	0.01025920	0.01025921
0.5	0.01345682	0.01287331	0.01282383	0.01282388	0.01282388
0.6	0.01613521	0.01544610	0.01538696	0.01538702	0.01538702
0.7	0.01879154	0.01799928	0.01793019	0.01793026	0.01793027
0.8	0.02122422	0.02029882	0.02021648	0.02021656	0.02021657
0.9	0.02070041	0.01952431	0.01941856	0.01941867	0.01941868

2.3 High-Order Numerical Solution of the Unsteady Burgers' Equation

One version of two-dimensional Burgers' equation is the unsteady Burgers' equation given by [31]

$$\frac{\partial u}{\partial t} + u\left(\frac{\partial u}{\partial x} + \frac{\partial u}{\partial y}\right) = \nu \left(\frac{\partial^2 u}{\partial x^2} + \frac{\partial^2 u}{\partial y^2}\right), \tag{2.61}$$
$$(x, y) \in D = [a, b] \times [c, d], \quad t > 0.$$

Here $\nu = 1/R > 0$ is an arbitrary number and R is the Reynold's number.

The aim of the present section is to construct stable and high-order FDMs to solve a unsteady Burgers' equation with inhomogeneous Dirichlet boundary conditions. It is realized by the following four steps.

1. The reduction of the unsteady Burgers' equation to the one-dimensional Burgers' equation using the properties of the required solution of the original equation.
2. To solve the obtained heat equation with Robin boundary conditions on the uniform grids of the spatial and time intervals by means of an explicit FDM. This method has a sixth-order approximation in the space variable, and a third-order approximation in the time variable, except boundary points of the spatial variable. We additionally used the fourth/sixth-order finite difference approximations for the Robin boundary conditions.
3. To find a numerical solution of the one-dimensional Burgers' equation by means of the numerical solution calculated in previous step of the heat equation. The obtained numerical solution has the same orders approximations in the space and time variables as numerical solution of the heat equation.

4. To find a numerical solution of the unsteady Burgers' equation by means of the numerical solution calculated in previous step of the one-dimensional Burgers' equation.

2.3.1   Reduction of the Unsteady Burgers' Equation to the One-Dimensional Burgers' Equation

It is easy to show that by linear transformation of independent variables

$$z = x + y, \quad s = x - y, \quad \bar{t} = 2t, \tag{2.62}$$

the Eq. (2.61) is reduced to the following equation:

$$\frac{\partial u}{\partial \bar{t}} + u \frac{\partial u}{\partial z} = \nu \left(\frac{\partial^2 u}{\partial z^2} + \frac{\partial^2 u}{\partial s^2} \right). \tag{2.63}$$

Note that the rectangular region D by the transformation (2.62) leads to interval $A = a + c \le z \le b + d = B$.

If the solution $u(z, s, t)$ depends only on variables s and t, the Eq. (2.63) leads to the heat equation

$$\frac{\partial u}{\partial \bar{t}} = \nu \frac{\partial^2 u}{\partial s^2}. \tag{2.64}$$

Also, if the solution $u(z, s, \bar{t})$ depends only on variables z and $\bar{t}$, i.e., $u(z, s, t) \equiv u(z, \bar{t})$, the Eq. (2.63) is reduced to the one-dimensional Burgers' equation

$$\frac{\partial u}{\partial \bar{t}} + u \frac{\partial u}{\partial z} = \nu \frac{\partial^2 u}{\partial z^2}, \tag{2.65}$$

with an initial condition

$$u(z, 0) = g(z, 0), \quad z \in (A, B). \tag{2.66}$$

The boundary conditions for Eq. (2.65) are defined by

$$u(A, \bar{t}) = g(A, \bar{t}), \quad u(B, \bar{t}) = g(B, \bar{t}). \tag{2.67}$$

There are many solutions of Eq. (2.61) that depend only on z and t [7–9, 32, 33]. As a examples, we present here some of these solutions

$$u(x, y, t) = \begin{cases} \left(1 + \exp \left(\frac{x+y-t}{2\nu} \right) \right)^{-1}, \\ \frac{1}{2} - \tanh \left(\frac{x+y-t}{2\nu} \right), \\ \frac{1}{2} - \coth \left(\frac{x+y+1-t}{2\nu} \right). \end{cases} \tag{2.68}$$

The main advantage of our approach is to reduce the two-dimensional unsteady Burgers' equation to one-dimensional one. This allows us to use known high-order numerical methods and to save computing time and memory of computers as compared to other direct methods [7–9] for solving two-dimensional unsteady Burgers' equation (2.61).

Here and below we consider only Eqs. (2.65)–(2.67). It is well known that, by the Hopf-Cole transformation

$$u(z, \bar{t}) = -2v \frac{1}{\theta(z, \bar{t})} \frac{\partial \theta(z, \bar{t})}{\partial z}, \tag{2.69}$$

the Burgers' equation (2.65) is reduced to the heat equation

$$\frac{\partial \theta(z, \bar{t})}{\partial \bar{t}} = v \frac{\partial^2 \theta(z, \bar{t})}{\partial z^2}. \tag{2.70}$$

By (2.69) the initial condition (2.66) and boundary conditions (2.67) lead to

$$\theta(z, 0) = \exp\left(-\frac{1}{2v} \int_A^z g(\xi, 0) d\xi\right), \tag{2.71}$$

and

$$\frac{\partial \theta(z, \bar{t})}{\partial z} + \frac{1}{2v} g(z, \bar{t}) \theta(z, \bar{t}) = 0 \quad \text{at} \quad z = A, B, \tag{2.72}$$

respectively. Thus, the Eqs. (2.65)–(2.67) are fully converted to problem (2.70)–(2.72).

2.3.2   Numerical Solution of the Heat Equation

We suppose that the solution of the heat problem defined by Eqs. (2.70)–(2.72) is a sufficiently smooth function with respect to z and $\bar{t}$. Then the heat problem can be solved by using high-order explicit methods proposed in the Sect. 2.2.1

$$\theta_i^{n+1} = \frac{\beta - \gamma}{\beta + \gamma} \theta_i^{n-1} + \frac{\beta \gamma}{\beta + \gamma} \left(\theta_{i-1}^n - 2\theta_i^n + \theta_{i+1}^n\right) + \frac{2\gamma}{\beta + \gamma} \theta_i^n, \tag{2.73}$$

$$\gamma = \frac{2\tau v}{h^2}, \quad h = \frac{B - A}{N}, \quad i = 1, \ldots, N - 1, \quad n = 1, 2, \ldots.$$

Here θ_i^n is the approximate solution at the mesh points $(z_i = ih, \bar{t}_n = n\tau)$, where h is a spatial step, τ is a time step, and N is the number of partition of the interval $[A, B]$. The method (2.73) is stable and its truncation error is of the order of $O(\tau^3 + h^6)$ provided that (2.12).

Note that, in the Sect. 2.2.1 the method (2.73) is used for the heat equation with Dirichlet boundary conditions. It is needed to adopt this method for equation (2.70) with Robin boundary conditions (2.72).

It should be mentioned that the method (2.73) is a three-level one in time. Hence, in order to find θ_i^n at level two, it requires values θ_i^n at level $n = 0, 1$, i.e., θ_i^0 and θ_i^1. Using the Taylor expansion of $\theta(z, \tau)$ at point $(z, 0)$ and Eq. (2.70), we obtain

$$\theta(z, \tau) = \theta(z, 0) + \nu \frac{\partial^2 \theta(z, 0)}{\partial z^2} \tau + \frac{\nu^2}{2} \frac{\partial^4 \theta(z, 0)}{\partial z^4} \tau^2 + \cdots . \qquad (2.74)$$

From the initial condition (2.71) and Taylor expansion (2.74), we find θ_i^1 with the accuracy $O(\tau^3)$

$$\theta_i^1 = \left(1 + \nu \tau F_1(z_i) + \frac{\nu^2 \tau^2}{2} F_2(z_i)\right) \exp\left(-\frac{1}{2\nu} \int_A^{z_i} g(\xi, 0) d\xi\right), \qquad (2.75)$$

where $i = 0, \ldots, N$, and

$$F_1(z) = -\frac{g_z'(z, 0)}{2\nu} + \frac{g^2(z, 0)}{4\nu^2}, \qquad (2.76)$$

$$F_2(z) = -\frac{g_z'''(z, 0)}{2\nu} + \frac{4g(z, 0)g_z''(z, 0) + 3(g_z'(z, 0))^2}{4\nu^2}$$

$$-\frac{3g^2(z, 0)g_z'(z, 0)}{4\nu^3} + \frac{g^4(z, 0)}{16\nu^4}.$$

From the Robin boundary conditions (2.72) using the asymmetric fourth-order and sixth-order finite difference approximations of the first spatial derivative [34]:

$$\frac{\partial \theta(z, \bar{t})}{\partial z}\bigg|_{z=A} = \frac{-25\theta_0^n + 48\theta_1^n - 36\theta_2^n + 16\theta_3^n - 3\theta_4^n}{12h} + O(h^4),$$

$$\frac{\partial \theta(z, \bar{t})}{\partial z}\bigg|_{z=B} = \frac{25\theta_N^n - 48\theta_{N-1}^n + 36\theta_{N-2}^n - 16\theta_{N-3}^n + 3\theta_{N-4}^n}{12h} + O(h^4), \qquad (2.77)$$

$$\frac{\partial \theta(z, \bar{t})}{\partial z}\bigg|_{z=A} = \frac{-147\theta_0^n + 360\theta_1^n - 450\theta_2^n + 400\theta_3^n - 225\theta_4^n}{60h}$$

$$+\frac{72\theta_5^n - 10\theta_6^n}{60h} + O(h^6),$$

$$\frac{\partial \theta(z, \bar{t})}{\partial z}\bigg|_{z=B} = \frac{147\theta_N^n - 360\theta_{N-1}^n + 450\theta_{N-2}^n - 400\theta_{N-3}^n + 225\theta_{N-4}^n}{60h}$$

$$+\frac{-72\theta_{N-5}^n + 10\theta_{N-6}^n}{60h} + O(h^6),$$

we find the fourth-order and sixth-order approximations of θ_0^n and θ_N^n

$$\theta_0^n = \frac{48\theta_1^n - 36\theta_2^n + 16\theta_3^n - 3\theta_4^n}{25 - \frac{6h}{\nu}g(A, \bar{t}_n)}, \tag{2.78}$$

$$\theta_N^n = \frac{48\theta_{N-1}^n - 36\theta_{N-2}^n + 16\theta_{N-3}^n - 3\theta_{N-4}^n}{25 + \frac{6h}{\nu}g(B, \bar{t}_n)},$$

and

$$\theta_0^n = \frac{360\theta_1^n - 450\theta_2^n + 400\theta_3^n - 225\theta_4^n + 72\theta_5^n - 10\theta_6^n}{147 - \frac{30h}{\nu}g(A, \bar{t}_n)}, \tag{2.79}$$

$$\theta_N^n = \frac{360\theta_{N-1}^n - 450\theta_{N-2}^n + 400\theta_{N-3}^n - 225\theta_{N-4}^n + 72\theta_{N-5}^n - 10\theta_{N-6}^n}{147 + \frac{30h}{\nu}g(B, \bar{t}_n)},$$

respectively. Thus, we find θ_i^n for $i = 0, \ldots, N$ by the formulas (2.73), (2.78) or (2.73), (2.79).

2.3.3 High-Order Finite Difference Methods for Solution of One-Dimensional Burgers' Equation

The high-order FDMs, which presented in the Sect. 2.2.1, are applied for solving the Eqs. (2.65)–(2.67). For convenience, we recall them shortly. First, we consider the fourth-order FDM (2.25) with boundary conditions

$$v_0^n = \theta_0^n\, g(A, \bar{t}_n), \quad v_N^n = \theta_N^n\, g(B, \bar{t}_n). \tag{2.80}$$

Here $v_i^n = \theta_i^n w_i^n$ and $w_i^n \equiv w(z_i, \bar{t}_n)$ is an approximate solution of $u(z_i, \bar{t}_n)$.

The sixth-order FDM has the form (2.30). Of course, besides of (2.80), we need additionally two end conditions v_1^n and v_{N-1}^n in order to solve the system (2.30). The solution of penta-diagonal system (2.30) leads to two three-diagonal systems (2.40) and (2.39). Now, the required additionally two end conditions Z_1^n and Z_{N-1}^n, are obtained from (2.39) and (2.69) as

$$Z_1^n = -2\nu(\theta_z'(A, \bar{t}_n) + a\theta_z'(z_1, \bar{t}_n) + \theta_z'(z_2, \bar{t}_n)), \tag{2.81}$$

$$Z_{N-1}^n = -2\nu(\theta_z'(z_{N-2}, \bar{t}_n) + a\theta_z'(z_{N-1}, \bar{t}_n) + \theta_z'(B, \bar{t}_n)).$$

Using the Robin boundary conditions (2.72) and the asymmetric sixth-order finite difference approximations of the first spatial derivative [34], we can find the needed terms $\theta_z'(A, \bar{t}_n)$, $\theta_z'(z_1, \bar{t}_n)$, $\theta_z'(z_2, \bar{t}_n)$, $\theta_z'(z_{N-2}, \bar{t}_n)$, $\theta_z'(z_{N-1}, \bar{t}_n)$ and $\theta_z'(B, \bar{t}_n)$:

$$\theta'_z(A, \bar{t}_n) = -\frac{1}{2\nu} g(A, \bar{t}_n) \theta^n_0,$$

$$\theta'_z(z_1, \bar{t}_n) = \frac{-10\theta^n_0 - 77\theta^n_1 + 150\theta^n_2 - 100\theta^n_3 + 50\theta^n_4 - 15\theta^n_5 + 2\theta^n_6}{60h},$$

$$\theta'_z(z_2, \bar{t}_n) = \frac{2\theta^n_0 - 24\theta^n_1 - 35\theta^n_2 + 80\theta^n_3 - 30\theta^n_4 + 8\theta^n_5 - \theta^n_6}{60h}, \tag{2.82}$$

$$\theta'_z(z_{N-2}, \bar{t}_n) = \frac{\theta^n_{N-6} - 8\theta^n_{N-5} + 30\theta^n_{N-4} - 80\theta^n_{N-3} + 35\theta^n_{N-2} + 24\theta^n_{N-1} - 2\theta^n_N}{60h},$$

$$\theta'_z(z_{N-1}, \bar{t}_n) = \frac{-2\theta^n_{N-6} + 15\theta^n_{N-5} - 50\theta^n_{N-4} + 100\theta^n_{N-3} - 150\theta^n_{N-2}}{60h}$$

$$+ \frac{77\theta^n_{N-1} + 10\theta^n_N}{60h},$$

$$\theta'_z(B, \bar{t}_n) = -\frac{1}{2\nu} g(B, \bar{t}_n) \theta^n_N.$$

Substituting (2.82) into (2.81), we find Z^n_1 and Z^n_{N-1} with order $O(h^6)$. Hence, the system of Eqs. (2.40), (2.81) is solved by the efficient elimination method [30]. After them, one can solve (2.39) with

$$v^n_0 = -2\nu \, \theta'_z(A, \bar{t}_n), \quad v^n_N = -2\nu \, \theta'_z(B, \bar{t}_n). \tag{2.83}$$

Thus, we obtain the numerical solution $w^n_i = v^n_i / \theta^n_i$ of the Burgers' equation (2.65)–(2.67) with a higher accuracy provided that the solution θ^n_i of heat equation (2.70)–(2.72) is found with a higher accuracy. The approximate values of solution of Eq. (2.61) with initial condition (2.66) and boundary condition (2.67) are found by formula

$$u^n_{kj} = w^n_i, \quad i = 0, \dots, N, \tag{2.84}$$

where $u^n_{kj} = u(x_k, y_j, t_n)$, $x_k + y_j = z_i$, $a \le x_k \le b$, $c \le y_j \le d$ and $t_n = \bar{t}_n / 2$.

2.3.4 Numerical Results

In this section we demonstrate the accuracy of the fourth-order and sixth-order FDMs proposed, respectively, by solving three exact solvable unsteady Burgers' equation (2.61) and compare the numerical results $w(z, T)$ with the exact results $u(z, T)$. The maximum absolute error of the solution is defined by

$$\|e\|_{\infty,h} = \max_{0 \le i \le N} |w(z_i, T) - u(z_i, T)|. \tag{2.85}$$

Convergence of the proposed methods is reviewed by computation of the Runge coefficient

$$r_h = \frac{\|e\|_{\infty,h}}{\|e\|_{\infty,h/2}}. \tag{2.86}$$

The all computations are performed using a Fortran program with quadruple-precision arithmetics.

We consider the following exact solutions of Eqs. (2.65):

$$u(z, \bar{t}) = \begin{cases} \left(1 + \exp\left(\frac{z}{2\nu} - \frac{\bar{t}}{4\nu}\right)\right)^{-1}, \\ \frac{1}{2} - \tanh\left(\frac{z}{2\nu} - \frac{\bar{t}}{4\nu}\right), \\ \frac{1}{2} - \coth\left(\frac{z+1}{2\nu} - \frac{\bar{t}}{4\nu}\right). \end{cases} \tag{2.87}$$

Note that the second and third solutions called the kink and travelling wave solutions of the Burgers' equation [33], respectively. In all the examples the initial and boundary conditions are taken from the exact solutions. It is easy to show that the corresponding heat problems (2.70)–(2.72) are also exact solvable, and normalized exact solutions with condition $\theta(0, 0) = 1$ can be expressed as

$$\theta(z, \bar{t}) = \begin{cases} \frac{1}{2}\left(1 + \exp\left(-\frac{z}{2\nu} + \frac{\bar{t}}{4\nu}\right)\right), & \textbf{Example 2.3,} \\ \exp\left(-\frac{z}{4\nu} + \frac{5\bar{t}}{16\nu}\right)\cosh\left(\frac{z}{2\nu} - \frac{\bar{t}}{4\nu}\right), & \textbf{Example 2.4,} \\ \exp\left(-\frac{z}{4\nu} + \frac{5\bar{t}}{16\nu}\right)\frac{\sinh\left(\frac{z+1}{2\nu} - \frac{\bar{t}}{4\nu}\right)}{\sinh\left(\frac{1}{2\nu}\right)}, & \textbf{Example 2.5.} \end{cases} \tag{2.88}$$

The maximum absolute errors $\|e\|_{\infty,h}$ and the Runge coefficients r_h for the **Examples 2.3–2.5** are presented in Tables 2.12, 2.13 and 2.14. They are consistent with the theoretical expectations of $O(h^4)$ and $O(h^6)$. Also we see that for small number ν required large N to obtain high-order numerical solutions. Note that from Tables 2.12, 2.13 and 2.14 we observed a slow convergence of the Runge coefficients r_h to the theoretical expectations (especially at small values of ν). This fact is a consequence of the application of the asymmetric approximations (2.78), (2.79) and (2.82) in a vicinity of the boundary points and due to the presence of large gradient in the solution. The results of numerical experiments demonstrate the expected accuracy and convergence order of proposed methods.

2.4 High-Order Numerical Solution of the Two-Dimensional Coupled Burgers' Equation

The two-dimensional coupled Burgers' equation (TDCBE) is given by [35]

$$\frac{\partial u}{\partial t} + u\frac{\partial u}{\partial x} + v\frac{\partial u}{\partial y} = \nu\left(\frac{\partial^2 u}{\partial x^2} + \frac{\partial^2 u}{\partial y^2}\right), \tag{2.89}$$

$$\frac{\partial v}{\partial t} + u\frac{\partial v}{\partial x} + v\frac{\partial v}{\partial y} = \nu\left(\frac{\partial^2 v}{\partial x^2} + \frac{\partial^2 v}{\partial y^2}\right), \tag{2.90}$$

Table 2.12 The maximum absolute error $\|e\|_{\infty,h}$ and the Runge coefficient r_h for **Example** 2.3. The factor x in the brackets denotes 10^x

	N	Fourth-order FDM		Sixth-order FDM	
		$\|e\|_{\infty,h}$	r_h	$\|e\|_{\infty,h}$	r_h
$v = 1$	20	6.81630(-07)		1.09301(-09)	
$T = 1/\sqrt{15}$	40	4.60638(-08)	14.80	1.89602(-11)	57.65
	80	2.99137(-09)	15.40	3.11914(-13)	60.79
	160	1.90541(-10)	15.70	4.99992(-15)	62.38
	320	1.20218(-11)	15.85	7.91258(-17)	63.19
	640	7.54910(-13)	15.92	1.24423(-18)	63.59
	1280	4.72930(-14)	15.96	1.95030(-20)	63.79
$v = 0.1$	20	4.09922(-03)		4.22744(-04)	
$T = 1/\sqrt{15}$	40	4.03057(-04)	10.17	1.33297(-05)	31.71
	80	3.15596(-05)	12.77	2.94802(-07)	45.21
	160	2.20669(-06)	14.30	5.47935(-09)	53.80
	320	1.45857(-07)	15.13	9.33799(-11)	58.67
	640	9.37449(-09)	15.56	1.52382(-12)	61.28
	1280	5.94147(-10)	15.78	2.43325(-14)	62.62
$v = 0.01$	160	1.92093(-01)		2.93743(-02)	
$T = 10/\sqrt{15}$	320	1.85362(-02)	10.36	9.64375(-04)	30.45
	640	1.48557(-03)	12.48	2.23559(-05)	43.13
	1280	1.05437(-04)	14.09	4.27027(-07)	52.35
	2560	7.03016(-06)	14.99	7.38644(-09)	57.81

subject to the initial conditions

$$u(x, y, 0) = \varphi_1(x, y), \quad (x, y) \in \Omega, \tag{2.91}$$
$$v(x, y, 0) = \varphi_2(x, y), \quad (x, y) \in \Omega, \tag{2.92}$$

and Dirichlet boundary conditions

$$u(x, y, t) = \zeta(x, y, t), \quad (x, y) \in \partial\Omega, \quad t > 0, \tag{2.93}$$
$$v(x, y, t) = \xi(x, y, t), \quad (x, y) \in \partial\Omega, \quad t > 0. \tag{2.94}$$

Here $\Omega = \{(x, y) : a \leq x \leq b, c \leq y \leq d\}$ is the computational domain, and $\partial\Omega$ is its boundary; $u(x, y, t)$ and $v(x, y, t)$ are the velocity components to be determined; $\varphi_1(x, y)$, $\varphi_2(x, y)$, $\zeta(x, y, t)$ and $\xi(x, y, t)$ are known functions; v^{-1} is the Reynolds number.

Table 2.13 The same as in Table 2.12, but for **Example** 2.4

	N	Fourth-order FDM		Sixth-order FDM	
		$\|e\|_{\infty,h}$	r_h	$\|e\|_{\infty,h}$	r_h
$\nu = 1$	320	1.30462(-06)		1.85387(-09)	
$T = 1/\sqrt{15}$	640	8.44299(-08)	15.45	3.06840(-11)	60.41
	1280	5.36962(-09)	15.72	4.93466(-13)	62.18
	2560	3.38538(-10)	15.86	7.82248(-15)	63.08
	5120	2.12510(-11)	15.93	1.23111(-16)	63.53
	10240	1.33108(-12)	15.96	1.93057(-18)	63.76
$\nu = 0.01$	640	3.63989(-03)		1.14887(-04)	
$T = 1/\sqrt{15}$	1280	2.74537(-04)	13.25	2.41846(-06)	47.50
	2560	1.89142(-05)	14.51	4.40048(-08)	54.95
	5120	1.24096(-06)	15.24	7.42060(-10)	59.30
	10240	7.94748(-08)	15.61	1.20472(-11)	61.59

Table 2.14 The same as in Table 2.12, but for **Example** 2.5

	N	Fourth-order FDM		Sixth-order FDM	
		$\|e\|_{\infty,h}$	r_h	$\|e\|_{\infty,h}$	r_h
$\nu = 1$	320	7.53462(-09)		1.58377(-12)	
$T = 1/\sqrt{15}$	640	4.74283(-10)	15.88	2.52478(-14)	62.72
	1280	2.97497(-11)	15.94	3.98712(-16)	63.32
	2560	1.86273(-12)	15.97	6.26406(-18)	63.65
	5120	1.16526(-13)	15.98	9.81480(-20)	63.82
	10240	7.28624(-15)	15.99	1.53570(-21)	63.91
$\nu = 0.01$	320	7.24049(-05)		1.37914(-06)	
$T = 1/\sqrt{15}$	640	4.55170(-06)	15.90	2.01708(-08)	68.37
	1280	2.85640(-07)	15.93	3.05365(-10)	66.05
	2560	1.78938(-08)	15.96	4.69822(-12)	64.99
	5120	1.11973(-09)	15.98	7.28515(-14)	64.49
	10240	7.00275(-11)	15.98	1.13399(-15)	64.24

Using the Hopf-Cole transformations [15, 36]

$$u(x, y, t) = -2v\frac{1}{\theta(x, y, t)}\frac{\partial\theta(x, y, t)}{\partial x}, \qquad (2.95)$$

$$v(x, y, t) = -2v\frac{1}{\theta(x, y, t)}\frac{\partial\theta(x, y, t)}{\partial y}, \qquad (2.96)$$

the Eqs. (2.89) and (2.90) are reduced to the two-dimensional heat equation (TDHE)

$$\frac{\partial\theta(x, y, t)}{\partial t} - v\left(\frac{\partial^2\theta(x, y, t)}{\partial x^2} + \frac{\partial^2\theta(x, y, t)}{\partial y^2}\right) - C(t)\theta(x, y, t) = 0, \qquad (2.97)$$

where $C(t)$ is an arbitrary function depending on t only.

Theorem 2.2 *[15] Let $\theta(x, y, t)$ be the solution of the Eq. (2.97), the functions $u(x, y, t)$ and $v(x, y, t)$ are defined in Eqs. (2.95) and (2.96). Then $u(x, y, t)$ and $v(x, y, t)$ are independent of $C(t)$.*

By the above theorem, we can choose $C(t) = 0$, and Eq. (2.97) is simplified to

$$\frac{\partial\theta(x, y, t)}{\partial t} = v\left(\frac{\partial^2\theta(x, y, t)}{\partial x^2} + \frac{\partial^2\theta(x, y, t)}{\partial y^2}\right). \qquad (2.98)$$

The initial conditions (2.91), (2.92) and boundary conditions (2.93), (2.94) lead to

$$\theta(x, y, 0) = \Phi(x, y), \qquad (2.99)$$

$$\left.\frac{\partial\theta(x, y, t)}{\partial x}\right|_{x=a} + \frac{\zeta(a, y, t)}{2v}\theta(a, y, t) = 0, \qquad (2.100)$$

$$\left.\frac{\partial\theta(x, y, t)}{\partial x}\right|_{x=b} + \frac{\zeta(b, y, t)}{2v}\theta(b, y, t) = 0, \qquad (2.101)$$

$$\left.\frac{\partial\theta(x, y, t)}{\partial y}\right|_{y=c} + \frac{\xi(x, c, t)}{2v}\theta(x, c, t) = 0, \qquad (2.102)$$

$$\left.\frac{\partial\theta(x, y, t)}{\partial y}\right|_{y=d} + \frac{\xi(x, d, t)}{2v}\theta(x, d, t) = 0, \qquad (2.103)$$

respectively. Here the function $\Phi(x, y)$ has the form [15]

$$\Phi(x, y) = \exp\left(-\frac{1}{2v}\int_a^x \varphi_1(s, y)ds - \frac{1}{2v}\int_c^y \varphi_2(a, s)ds\right). \qquad (2.104)$$

Thus, the TDCBE (2.89)–(2.94) are fully reduced to TDHE (2.98) with the initial and boundary conditions (2.99)–(2.103).

2.4.1   The Fourth-Order Explicit Finite Difference Method

For the TDHE (2.98)–(2.103), we consider the following eleven-points explicit FDM:

$$
F(\theta) \equiv \frac{A\,\theta_{i,j}^{n+1} - (A+B)\,\theta_{i,j}^{n} + B\,\theta_{i,j}^{n-1}}{\tau}
$$

$$
-C\,\frac{\theta_{i-1,j}^{n} - 2\theta_{i,j}^{n} + \theta_{i+1,j}^{n}}{h_x^2} - C\,\frac{\theta_{i,j-1}^{n} - 2\theta_{i,j}^{n} + \theta_{i,j+1}^{n}}{h_y^2}
$$

$$
-D\left(\frac{1}{h_x^2} + \frac{1}{h_y^2}\right)(\theta_{i-1,j-1}^{n} + \theta_{i-1,j+1}^{n} + \theta_{i+1,j-1}^{n} + \theta_{i+1,j+1}^{n} - 4\theta_{i,j}^{n}) \qquad (2.105)
$$

$$
i = 1,\dots,N-1,\quad j = 1,\dots,M-1,\quad h_x = \frac{b-a}{N},\quad h_y = \frac{d-c}{M},\quad n = 1,2,\dots.
$$

Here $\theta_{i,j}^{n}$ is the approximate solution at the mesh point $(x_i = ih_x,\ y_j = jh_y,\ t_n = n\tau)$, where h_x and h_y are spatial steps by x and y, τ is a time step, A, B, C and D are unknown coefficients. We suppose that the solution of Eqs. (2.98)–(2.103) is a sufficiently smooth function with respect to x, y and t. Let

$$
z_{i,j}^{n} = \theta(x_i, y_j, t_n) - \theta_{i,j}^{n} \qquad (2.106)
$$

be the error function. In this term the method (2.105) has the form

$$
F(z) = \psi_{i,j}^{n}, \qquad (2.107)
$$

where $\psi_{i,j}^{n}$ is an approximation error and

$$
\psi_{i,j}^{n} = F(\theta). \qquad (2.108)
$$

We suppose that the solution of Eqs. (2.98)–(2.103) is a sufficiently smooth function with respect to x, y and t. Using the Taylor expansions of $\theta_{i,j}^{n+1}$, $\theta_{i\pm1,j}^{n}$, $\theta_{i,j\pm1}^{n}$ and $\theta_{i\pm1,j\pm1}^{n}$ at the point (x_i, y_j, t_n), and an identity

$$
\frac{\partial^m \theta(x, y, t)}{\partial t^m} = v^m \left(\frac{\partial^2}{\partial x^2} + \frac{\partial^2}{\partial y^2}\right)^m \theta(x, y, t), \quad m \geq 0, \qquad (2.109)
$$

we have

$$
\psi_{i,j}^{n} = \left\{ \left[(A-B)v - C - 2D\left(1 + \frac{h_x^2}{h_y^2}\right) \right] \frac{\partial^2 \theta(x, y, t)}{\partial x^2} \right.
$$

$$
+ \left[(A-B)v - C - 2D\left(1 + \frac{h_y^2}{h_x^2}\right) \right] \frac{\partial^2 \theta(x, y, t)}{\partial y^2}
$$

$$
+ \left[(A+B)\frac{\tau v^2}{2} - C\frac{h_x^2}{12} - D\left(\frac{h_x^2}{6} + \frac{h_x^4}{6h_y^2}\right) \right] \frac{\partial^4 \theta(x, y, t)}{\partial x^4}
$$

$$+\left[(A+B)\frac{\tau v^2}{2} - C\frac{h_y^2}{12} - D\left(\frac{h_y^2}{6} + \frac{h_y^4}{6h_x^2}\right)\right]\frac{\partial^4\theta(x,y,t)}{\partial y^4}$$

$$+\left[(A+B)\tau v^2 - D(h_x^2 + h_y^2)\right]\frac{\partial^4\theta(x,y,t)}{\partial x^2\partial y^2}\Bigg\}\Bigg|_{x-x_i,y-y_j,t-t_n}$$

$$+O\left(h_x^4 + h_y^4 + h_x^2 h_y^2 + \frac{h_x^6}{h_y^2} + \frac{h_y^6}{h_x^2}\right). \tag{2.110}$$

Equating the coefficients of the partial derivatives to zero in (2.110), we obtain following system of equations

$$\begin{cases} (A-B)v = C + 2D\left(1 + \frac{h_x^2}{h_y^2}\right) = C + 2D\left(1 + \frac{h_y^2}{h_x^2}\right), \\ (A+B)\tau v^2 = C\frac{h_x^2}{6} + D\left(\frac{h_x^2}{3} + \frac{h_x^4}{3h_y^2}\right) = C\frac{h_y^2}{6} + D\left(\frac{h_y^2}{3} + \frac{h_y^4}{3h_x^2}\right), \\ (A+B)\tau v^2 = D(h_x^2 + h_y^2). \end{cases} \tag{2.111}$$

The above system has an unique solution if and only if $h = h_x = h_y$:

$$C = 8D = \frac{4A\alpha v}{1+3\alpha}, \quad B = A\frac{1-3\alpha}{1+3\alpha}, \tag{2.112}$$

where $\alpha = 2\tau v/h^2$. Using (2.112) and the Taylor expansions of the $\theta(x,y,t)$ at the point (x_i, y_j, t_n) we obtain

$$\psi_{i,j}^n = D\frac{h^4}{2}\left\{\left(\alpha^2 - \frac{1}{15}\right)\left(\frac{\partial^6\theta(x,y,t)}{\partial x^6} + \frac{\partial^6\theta(x,y,t)}{\partial y^6}\right)\right.$$

$$\left.+3\left(\alpha^2 - \frac{1}{9}\right)\left(\frac{\partial^6\theta(x,y,t)}{\partial x^4\partial y^2} + \frac{\partial^6\theta(x,y,t)}{\partial x^2\partial y^4}\right)\right\}\Bigg|_{x=x_i,y=y_j,t=t_n}$$

$$+O(\tau^3 + h^6). \tag{2.113}$$

One can see that, the condition (2.12)

$$\alpha = \gamma = \frac{2\tau v}{h^2} = \sqrt{\frac{1}{15}} \tag{2.114}$$

does not improve the order of the method (2.105), i.e., the truncation error of the method (2.105) is of the order of $\psi_{i,j}^n = O(\tau^2 + h^4)$ for any α. If $A = 1$, the method (2.105) is simplified to the canonical form:

$$\theta_{i,j}^{n+1} = -\frac{1-3\alpha}{1+3\alpha}\theta_{i,j}^{n-1} + \frac{2-10\alpha^2}{1+3\alpha}\theta_{i,j}^{n}$$

$$+\frac{2\alpha^2}{1+3\alpha}\left(\theta_{i-1,j}^{n} + \theta_{i+1,j}^{n} + \theta_{i,j-1}^{n} + \theta_{i,j+1}^{n}\right. \tag{2.115}$$

$$\left.+\frac{1}{4}\left(\theta_{i-1,j-1}^{n} + \theta_{i-1,j+1}^{n} + \theta_{i+1,j-1}^{n} + \theta_{i+1,j+1}^{n}\right)\right).$$

For finding the stability condition of the method (2.115), we seek the partial solution in the form:

$$\theta_{i,j}^{n} = q^n \exp(\iota i h\psi)\exp(\iota jh\phi). \tag{2.116}$$

From (2.115) we have

$$(1+3\alpha)q^2 + 2bq + 1 - 3\alpha = 0, \tag{2.117}$$

$$b = (9 - A)\alpha^2 - 1,$$

$$1 \le A = (\cos(h\psi) + 2)(\cos(h\phi) + 2) \le 9.$$

Using the conditions (2.16), we obtain

$$\begin{cases} \left|\frac{c}{a}\right| = \left|\frac{1-3\alpha}{1+3\alpha}\right| \le 1, \\ a + c = 2 \ge -2b = 2(1 - (9 - A)\alpha^2), \\ a + c = 2 \ge 2b = 2((9 - A)\alpha^2 - 1), \end{cases} \Rightarrow \begin{cases} 0 \le \alpha, \\ A \le 9, \\ \alpha \le \sqrt{\frac{2}{9-A}}. \end{cases} \tag{2.118}$$

The last inequality is true for any A, that $\alpha \le 1/3$ or

$$\tau \le \frac{h^2}{4\nu}. \tag{2.119}$$

The method (2.115) at $\alpha = 1/3$ (or $B = 0$) is a two-layer scheme in time, while at $\alpha \ne 1/3$ (or $B \ne 0$) is a three-layer one. Hence, if $\alpha \ne 1/3$, in order to find $\theta_{i,j}^2$ at level two it requires two values $\theta_{i,j}^0$ and $\theta_{i,j}^1$. Using the Taylor expansion of $\theta(x, y, t)$ at point $(x, y, 0)$ and Eq. (2.98) we obtain

$$\theta(x, y, t) = \theta(x, y, 0) + \tau\nu\left(\frac{\partial^2\theta(x, y, 0)}{\partial x^2} + \frac{\partial^2\theta(x, y, 0)}{\partial y^2}\right) + O(\tau^2). \tag{2.120}$$

From the initial condition (2.99) and Taylor expansion (2.120), we find $\theta_{i,j}^1$ with the accuracy $O(\tau^3)$

$$\theta_{i,j}^1 = \Phi(x_i, y_j) + \tau\nu\left(\frac{\partial^2\Phi(x, y)}{\partial x^2} + \frac{\partial^2\Phi(x, y)}{\partial y^2}\right)\Bigg|_{x=x_i, y=y_j}, \tag{2.121}$$

$$i = 0, \ldots, N, \quad j = 0, \ldots, M.$$

From the Robin boundary conditions (2.100)–(2.103) using the asymmetric fourth-order finite difference approximations of the first spatial derivative [34], we find $\theta_{0,j}^{n+1}, \theta_{N,j}^{n+1}, \theta_{i,0}^{n+1}$ and $\theta_{i,M}^{n+1}$

$$
\theta_{0,j}^{n+1} = \frac{48\theta_{1,j}^{n+1} - 36\theta_{2,j}^{n+1} + 16\theta_{3,j}^{n+1} - 3\theta_{4,j}^{n+1}}{25 - \frac{12h}{2\nu}\zeta(a, y_j, t_{n+1})}, \quad j = 1, \ldots, M - 1,
$$

$$
\theta_{N,j}^{n+1} = -\frac{3\theta_{N-4,j}^{n+1} - 16\theta_{N-3,j}^{n+1} + 36\theta_{N-2,j}^{n+1} - 48\theta_{N-1,j}^{n+1}}{25 + \frac{12h}{2\nu}\zeta(b, y_j, t_{n+1})},
$$

$$
\theta_{i,0}^{n+1} = \frac{48\theta_{i,1}^{n+1} - 36\theta_{i,2}^{n+1} + 16\theta_{i,3}^{n+1} - 3\theta_{i,4}^{n+1}}{25 - \frac{12h}{2\nu}\xi(x_i, c, t_{n+1})}, \quad i = 1, \ldots, N - 1,
$$

$$
\theta_{i,M}^{n+1} = -\frac{3\theta_{i,M-4}^{n+1} - 16\theta_{i,M-3}^{n+1} + 36\theta_{i,M-2}^{n+1} - 48\theta_{i,M-1}^{n+1}}{25 + \frac{12h}{2\nu}\xi(x_i, d, t_{n+1})}.
$$

(2.122)

Now we needed to calculate values of the vertex points $\theta_{0,0}^{n+1}, \theta_{N,0}^{n+1}, \theta_{0,M}^{n+1}$ and $\theta_{N,M}^{n+1}$. Each value of these points can be calculate using the boundary conditions (2.100)–(2.103) and similar formula to (2.122) by direction x or y or middle value of the values by the both directions. Below we presented formulas which used only the boundary conditions (2.100), (2.101):

$$
\theta_{0,0}^{n+1} = \frac{48\theta_{1,0}^{n+1} - 36\theta_{2,0}^{n+1} + 16\theta_{3,0}^{n+1} - 3\theta_{4,0}^{n+1}}{25 - \frac{12h}{2\nu}\zeta(a, c, t_{n+1})},
$$

$$
\theta_{N,0}^{n+1} = -\frac{3\theta_{N-4,0}^{n+1} - 16\theta_{N-3,0}^{n+1} + 36\theta_{N-2,0}^{n+1} - 48\theta_{N-1,0}^{n+1}}{25 + \frac{12h}{2\nu}\zeta(b, c, t_{n+1})},
$$

$$
\theta_{0,M}^{n+1} = \frac{48\theta_{1,M}^{n+1} - 36\theta_{2,M}^{n+1} + 16\theta_{3,M}^{n+1} - 3\theta_{4,M}^{n+1}}{25 - \frac{12h}{2\nu}\zeta(a, d, t_{n+1})},
$$

$$
\theta_{N,M}^{n+1} = -\frac{3\theta_{N-4,M}^{n+1} - 16\theta_{N-3,M}^{n+1} + 36\theta_{N-2,M}^{n+1} - 48\theta_{N-1,M}^{n+1}}{25 + \frac{12h}{2\nu}\zeta(b, d, t_{n+1})}.
$$

(2.123)

Thus, we find $\theta_{i,j}^n$ for $i = 0, \ldots, N$ and $j = 0, \ldots, M$ by the formulas (2.115), (2.122) and (2.123). We used the following fourth-order FDM proposed in the Sect. 2.2.1:

$$
\theta_{i-1,j}^n u_{i-1,j}^n + 4\theta_{i,j}^n u_{i,j}^n + \theta_{i+1,j}^n u_{i+1,j}^n = -\frac{6\nu}{h}\left(\theta_{i+1,j}^n - \theta_{i-1,j}^n\right), \quad (2.124)
$$

$$
i = 1, \ldots, N - 1, \quad j = 0, \ldots, M,
$$

$$
\theta_{i,j-1}^n v_{i,j-1}^n + 4\theta_{i,j}^n v_{i,j}^n + \theta_{i,j+1}^n v_{i,j+1}^n = -\frac{6\nu}{h}\left(\theta_{i,j+1}^n - \theta_{i,j-1}^n\right), \quad (2.125)
$$

$$
j = 1, \ldots, M - 1, \quad i = 0, \ldots, N,
$$

with boundary conditions

$$u^n_{0,j} = \zeta(a, y_j, t_n), \quad u^n_{N,j} = \zeta(b, y_j, t_n), \quad j = 0, \ldots, M. \tag{2.126}$$

$$v^n_{i,0} = \xi(x_i, c, t_n), \quad v^n_{i,M} = \xi(x_i, d, t_n), \quad i = 0, \ldots, N. \tag{2.127}$$

Here $u^n_{i,j}$ and $v^n_{i,j}$ are the approximate solutions of $u(x_i, y_j, t_n)$ and $v(x_i, y_j, t_n)$, respectively.

2.4.2   Numerical Results

In this section we demonstrate the accuracy of the fourth-order FDM proposed, by solving three exact solvable TDCBE (2.89)–(2.94) and compare the numerical results $U(x, y, t)$, $V(x, y, t)$ with the exact results $u(x, y, t)$, $v(x, y, t)$, respectively.

To analyze the convergence of the proposed method we used the maximum absolute errors of the solutions $u(x, y, t)$ and $v(x, y, t)$

$$\|e_u\|_{\infty,h} = \max_{0 \le i \le N, 0 \le j \le M} |u(x_i, y_j, t) - U(x_i, y_j, t)|,$$

$$\|e_v\|_{\infty,h} = \max_{0 \le i \le N, 0 \le j \le M} |v(x_i, y_j, t) - V(x_i, y_j, t)|. \tag{2.128}$$

The order of convergence of the proposed method (or Runge coefficient) defined by the double-crowding spatial grids

$$\text{Order} = \log_2 \left(\frac{\|e_{u,v}\|_{\infty,h}}{\|e_{u,v}\|_{\infty,h/2}} \right). \tag{2.129}$$

The all computations are performed using a Fortran program with quadruple-precision arithmetics.

We considered the numerical solutions to Eqs. (2.89)–(2.94) for the following three test examples. The initial and boundary conditions (2.91), (2.92) and (2.93), (2.94) for $u(x, y, t)$ and $v(x, y, t)$ are taken from the analytical solutions. The computational domain is $\Omega = \{(x, y) : 0 \le x \le 1, \; 0 \le y \le 1\}$ for first two examples, and is $\Omega = \{(x, y) : 0 \le x \le \frac{1}{2}, \; 0 \le y \le 1\}$ for third example.

Example 2.6

$$u(x, y, t) = \frac{-2vy + 2v\pi \exp\left(-2v\pi^2 t\right)\left(\cos(\pi x) + \sin(\pi x)\right)\sin(\pi y)}{100 + xy + \exp\left(-2v\pi^2 t\right)\left(\cos(\pi x) - \sin(\pi x)\right)\sin(\pi y)},$$

$$v(x, y, t) = \frac{-2vx - 2v\pi \exp\left(-2v\pi^2 t\right)\left(\cos(\pi x) - \sin(\pi x)\right)\cos(\pi y)}{100 + xy + \exp\left(-2v\pi^2 t\right)\left(\cos(\pi x) - \sin(\pi x)\right)\sin(\pi y)}. \tag{2.130}$$

Table 2.15 The grid size, the maximum absolute error and the order of convergence for solution u and v at $\nu = \frac{1}{4000}$, $t = 1$, $\tau = 0.001$ for the **Example** 2.6. The factor x in the brackets denotes 10^x

Grid size	u-component error	Order	v-component error	Order
4×4	0.51817($-$06)		0.54102($-$06)	
8×8	0.35823($-$07)	3.85447	0.38269($-$07)	3.82142
16×16	0.21268($-$08)	4.07413	0.22378($-$08)	4.09605
32×32	0.15090($-$09)	3.81702	0.18396($-$09)	3.60462
64×64	0.13779($-$10)	3.45300	0.18156($-$10)	3.34084
128×128	0.11178($-$11)	3.62371	0.15328($-$11)	3.56623
256×256	0.78381($-$13)	3.83407	0.10967($-$12)	3.80487

We solve the TDHE (2.98), for which the exact solution is

$$\theta(x, y, t) = 100 + xy + \exp\left(-2\nu\pi^2 t\right)\left(\cos(\pi x) - \sin(\pi x)\right)\sin(\pi y). \quad (2.131)$$

Example 2.7

$$u(x, y, t) = \frac{3}{4} - \frac{1}{4}\left(1 + \exp\left(\frac{-4x + 4y - t}{32\nu}\right)\right)^{-1},$$

$$v(x, y, t) = \frac{3}{4} + \frac{1}{4}\left(1 + \exp\left(\frac{-4x + 4y - t}{32\nu}\right)\right)^{-1}. \quad (2.132)$$

We solve the TDHE (2.98), for which the exact solution is

$$\theta(x, y, t) = \frac{1}{2}\exp\left(-\frac{12x + 12y - 9t}{32\nu}\right)\left(1 + \exp\left(\frac{4x - 4y + t}{32\nu}\right)\right). \quad (2.133)$$

The solutions (2.132) are so-called shock solutions of the TDCBE. It is well known that one of the difficulties in solving Burgers' equations is that shock of the solution may occur after some time, even if the initial functions are smooth. When the characteristic curves of Burgers' equation cross, a shock of the solution occurs. A robust and high-order numerical algorithm should be able to capture the shock and the numerical solution should exhibit the correct physical behavior.

Table 2.16 The same as in Table 2.15, but for the Crank–Nicolson method [11]

Grid size	u−component error	Order	v-component error	Order
4×4	0.13726(−08)		0.77396(−09)	
8×8	0.45893(−09)	1.58062	0.35595(−09)	1.12060
16×16	0.12700(−09)	1.85341	0.11092(−09)	1.68205
32×32	0.32955(−10)	1.94630	0.30243(−10)	1.87494
64×64	0.82772(−11)	1.99329	0.77349(−11)	1.96716
128×128	0.19992(−11)	2.04972	0.18769(−11)	2.04301

Table 2.17 The grid size, the maximum absolute error and the order of convergence for solution u and v at $\nu = 0.0001$, $t = 1$, $\tau = 0.1$ for the **Example** 2.7. The factor x in the brackets denotes 10^x

Grid size	u-component error	Order	v-component error	Order
4×4	0.20974(+00)		0.14179(+01)	
8×8	0.70525(−02)	4.89436	0.33806(−01)	5.39031
16×16	0.33792(−03)	4.38338	0.18858(−02)	4.16402
32×32	0.22747(−04)	3.89294	0.11149(−03)	4.08023
64×64	0.15294(−05)	3.89466	0.70751(−05)	3.97798
128×128	0.99277(−07)	3.94533	0.46903(−06)	3.91501

Example 2.8

$$u(x, y, t) = -\frac{4\nu\pi \exp\left(-5\nu\pi^2 t\right) \cos(2\pi x) \sin(\pi y)}{2 + \exp\left(-5\nu\pi^2 t\right) \sin(2\pi x) \sin(\pi y)},$$

$$v(x, y, t) = -\frac{2\nu\pi \exp\left(-5\nu\pi^2 t\right) \sin(2\pi x) \cos(\pi y)}{2 + \exp\left(-5\nu\pi^2 t\right) \sin(2\pi x) \sin(\pi y)}. \tag{2.134}$$

We solve the TDHE (2.98), for which the exact solution is

$$\theta(x, y, t) = \frac{2 + \exp\left(-5\nu\pi^2 t\right) \sin(2\pi x) \sin(\pi y)}{2}. \tag{2.135}$$

The results in Table 2.15, 2.17 and 2.19 are calculated by the proposed fourth-order FDMs, while the results in Table 2.16 are obtained by using the Crank–Nicolson method

Table 2.18 The same as in Table 2.17, but for the implicit logarithmic FDM [14]

Grid size	u-component error	Order	v-component error	Order
4 × 4	0.29037(−01)		0.18517(−01)	
8 × 8	0.93349(−02)	1.6372	0.58959(−02)	1.6511
16 × 16	0.19106(−02)	2.2886	0.12013(−02)	2.2951
32 × 32	0.43163(−03)	2.1462	0.27225(−03)	2.1416
64 × 64	0.93810(−04)	2.2020	0.61495(−04)	2.1464

Table 2.19 The grid size, the maximum absolute error and the order of convergence for solution u and v at $\nu = 0.1$, $t = 1$, $\tau = 0.0001$ for the **Example** 2.8. The factor x in the brackets denotes 10^x

Grid size	u-component error	Order	v-component error	Order
4 × 8	0.66015(−04)		0.15269(−03)	
8 × 16	0.18974(−04)	1.79880	0.28857(−04)	2.40364
16 × 32	0.15249(−05)	3.63723	0.21678(−05)	3.73461
32 × 64	0.10129(−06)	3.91212	0.14175(−06)	3.93477
64 × 128	0.64380(−08)	3.97574	0.86112(−08)	4.04102

[11]. The results in Table 2.18 are obtained by the implicit logarithmic FDM [14]. The orders of convergence of the proposed methods are consistent with the theoretical expectations $O(h^4)$ for the proposed fourth-order FDMs, and $O(h^2)$ for the Crank–Nicolson method and implicit logarithmic FDM.

References

1. V. Ulziibayar, High-order finite-difference schemes for numerical solution of some partial differential equations. Doctoral thesis, Ulaanbaatar, Mongolia, 2014
2. S. Kutluay, A.R. Bahadir, A. Özdes, Numerical solution of one-dimensional Burgers equation: explicit and exact-explicit finite difference methods. J. Comput. Appl. Math. **103**, 251–261 (1999)
3. M.C. Kweyu, W.A. Manyonge, A. Koross, V. Ssemaganda, Numerical solutions of the Burgers' system in two dimensions under varied initial and boundary conditions. Appl. Math. Sci. **6**, 5603–5615 (2012)
4. K. Pandy, V. Lajja, K.V. Amit, On a finite-difference scheme for Burgers' equation. Appl. Math. Comput. **215**, 2208–2214 (2009)
5. R. Jiwari, A Haar wavelet quasilinearization approach for numerical simulation of Burgers' equation. Comput. Phys. Commun. **183**, 2413–2423 (2012)
6. G.W. Wei, D.S. Zhang, D.J. Kouri, D.K. Hoffman, Distributed approximating functional approach to Burgers' equation in one and two space dimensions. Comput. Phys. Commun. **111**, 93–109 (1998)

7. Y. Duan, R. Liu, Lattice Boltzmann model for 2D unsteady Burgers' equation. J. Comput. Appl. Math. **206**, 432–439 (2007)
8. S.F. Radvan, Comparison of higher-order accurate schemes for solving the 2D unsteady Burgers' equation. J. Comput. Appl. Math. **174**, 383–397 (2004)
9. S. Kutluay, N.M. Yagmurlu, The modified Bi-quintic B-splines for solving the 2D unsteady Burgers' equation. Eur. Inter. J. Sci. Tech. **1**, 23–39 (2012)
10. V. Ulziibayar, T. Zhanlav, O. Chuluunbaatar, Higher-order accurate numerical solution of Burgers' equation. Int. J. Math. Sci. **33**, 1374–1378 (2013)
11. V.K. Srivastava, M. Tamsir, U. Bhardwaj, Y.V.S.S. Sanyasiraju, Crank-Nicolson scheme for numerical solution of two-dimensional coupled Burgers' equation. Int. J. Sci. Eng. Res. **2**, 44–50 (2011)
12. A.R. Bahadir, A fully implicit finite-difference scheme for two-dimensional Burgers' equations. Appl. Math. Comput. **167**, 131–137 (2003)
13. M. Tamsir, V.K. Srivastava, A semi-implicit finite-difference approach for two dimensional coupled Burgers' equations. Int. J. Sci. Eng. Res. **2**, 46–51 (2011)
14. V.K. Srivastava, M.K. Awasthi, S. Singh, An implicit logarithmic finite-difference technique for two dimensional coupled viscous Burgers' equation. AIP Adv. **3**(122105), 1–9 (2013)
15. G. Zhao, X. Yu, R. Zhang, The new numerical method for solving the system of two-dimensional Burgers' equations. Comput. Math. Appl. **62**, 3279–3291 (2011)
16. W.Y. Liao, A fourth order finite-difference method for solving the system of two-dimensional Burgers' equations. Int. J. Numer. Methods Fluids **64**, 565–590 (2010)
17. K. Cleophas, N. Benjamin, W. John, Hybrid crank-nicolson-du fort and frankel (CN-DF) scheme for the numerical solution of the 2D coupled Burgers' system. Appl. Math. Sci. **8**, 2353–2361 (2014)
18. N.N. Kalitkin, *Numerical Methods* (Nauka, Moscow, 1986). ((in Russian))
19. C.-L. Zhu, R.-H. Wang, Numerical solution of Burgers' equation by cubic-B-spline quasi-interpolation. Appl. Math. Comput. **208**, 260–272 (2009)
20. C.-L. Zhu, W.-S. Kang, Numerical solution of Burgers-Fisher equation by cubic B-spline quasi-interpolation. Appl. Math. Comput. **216**, 2679–2686 (2010)
21. S. Abbasbandy, M.T. Darbishi, A numerical solution of Burgers' equation by time discretization of Adomain's decomposition method. Appl. Math. Comput. **170**, 95–102 (2005)
22. I. Dag, D. Irk, B. Saka, A numerical solution of the Burgers' equation using cubic B-splines. Appl. Math. Comput. **163**, 199–211 (2005)
23. H.N.A. Ismail, A.A. Abd Rabboh, A restrictive Padé approximation for the solution of the generalized Fisher and Burger-Fisher equations. Appl. Math. Comput. **154**, 203–210 (2004)
24. M. Javidi, Spectral collocation method for the solution of the generalized Burger-Fisher equation. Appl. Math. Comput. **176**, 345–352 (2006)
25. M. Lülsu, T. Özis, Numerical solution of Burgers' equation with restrictive Taylor approximations. Appl. Math. Comput. **171**, 1192–1200 (2005)
26. R.D. Richtmyer, K.W. Morton, *Difference Methods for Initial-Value Problems* (New York, 1967)
27. T. Zhanlav, Difference schemes with improved accuracy for one-dimensional heat equation. Applied Mathematic, Irkutsk, 149–153 (1978) (in Russian)
28. I.A. Hassanien, K.K. Sharma, H.A. Hosham, Fourth-order finite difference method for solving Burgers' equation. Appl. Math. Comput. **170**, 781–800 (2005)
29. T. Zhanlav, Z-folding and its applications. Mong. Math. J. **17**, 68–74 (2013)
30. A.A. Samarskii, A.V. Goolin, *Numerical Methods* (Nauka, Moscow, 1989). ((in Russian))
31. T. Zhanlav, O. Chuluunbaatar, V. Ulziibayar, Higher-order accurate numerical solution of unsteady Burgers' equation. Appl. Math. Comput. **250**, 701–707 (2015)

32. A.H. Khater, R.S. Temsah, M.M. Hassan, A Chebyshev spectral collocation method for solving Burgers'-type equations. J. Comput. Appl. Math. **222**, 333–350 (2008)
33. A.M. Wazwaz, Travelling wave solutions to (2+1)-dimensional nonlinear evolution equations. J. Nat. Sci. Math. **1**, 1–13 (2007)
34. B. Fornberg, Generation of finite difference formulas on arbitrarily spaced grids. Math. Comput. **51**, 699–706 (1988)
35. T. Zhanlav, O. Chuluunbaatar, V. Ulziibayar, Higher-order numerical solution of two-dimensional coupled Burgers' equation. Am. J. Comput. Math. **6**(02), 120–129 (2016)
36. C.A. Fletcher, Generating exact solutions of the two-dimensional Burgers' equations. Int. J. Numer. Methods Fluids **3**, 213–216 (1983)

References

High-Accuracy Finite Element Methods for Solution of Discrete Spectrum Problems

3

Abstract

A symbolic-numeric algorithm implemented in the Maple system for constructing multivariable Hermitian finite elements are presented. The basis functions of finite elements are high-order polynomials determined from a specially constructed set of values of the polynomials themselves, their partial derivatives and their derivatives along the normals to the boundaries of the finite elements. Such a choice of polynomials makes it possible to construct a piecewise polynomial basis continuous across the boundaries of the elements together with the derivatives up to a given order, which is used to solve elliptic boundary-value problems (BVPs) using the high-accuracy finite element method (FEM). The efficiency and accuracy order of the FEM, the algorithm and the program are demonstrated by test examples of exactly solvable Helmholtz problems on a triangle, square and four-dimensional hypercube, depending on the number of finite elements of the domain partition, the number of piecewise polynomial basis functions, and the dimension of eigenvectors of the corresponding algebraic eigenvalue problems. Comparison of lower parts spectra of quadrupole-octupole-vibrational collective model obtained by FEM and finite difference method (FDM) in solving the two dimensional BVP with the numerical tabular coefficients is also given.

3.1 Introduction

The study of mathematical models that describe tunneling and channeling of composite quantum systems through multidimensional barriers, photo-ionization and photo-absorption in molecular, atomic, nuclear, and quantum-dimensional semiconductor systems, requires high-accuracy efficient algorithms and programs for solving multidimensional BVPs [1–7].

© The Author(s), under exclusive license to Springer Nature Switzerland AG 2024
U. Vandandoo et al., *High-Order Finite Difference and Finite Element Methods for Solving Some Partial Differential Equations*, Synthesis Lectures on Engineering, Science, and Technology, https://doi.org/10.1007/978-3-031-44784-6_3

In this direction, using the variation-projection BVP formulation and the FEM with Lagrange interpolation elements [8–10], symbolic-numerical algorithms and programs are elaborated [2, 11–24]. This implementation of the FEM using Lagrange interpolation polynomials such that it preserves only the continuity of the solution itself in the course of its numerical approximation on a finite element grid. However, in the above class of problems, particularly, in quantum-dimensional semiconductor systems, the continuity should be preserved not only for the solution (wave function) itself, but also for the probability current [9]. The required continuity of the solution derivatives can be preserved in FEM numerical approximation using Hermite interpolation polynomials (HIPs) [25, 26]. The construction of such basis functions, referred to as HIPs, is impossible on an arbitrary mesh of nodes. It is one of the most important and difficult problems in the FEM and its applications in different fields, solved to date only for some particular cases [27–40].

High-order FEMs yield highly accurate solutions to BVPs due to their fast convergence. However, they are not currently used to solve multidimensional problems, since their implementation requires large resources. This obstacle is gradually removed with the progress in computational technology.

The cornerstone hindering factor in the implementation of FEMs is the calculation of integrals. It is well known [8] that, as a result of applying the p-th order FEM to the solution of the discrete spectrum problem for the elliptic (Schrödinger) equation, the eigenfunctions and the eigenvalues are determined with accuracies of the order $p + 1$ and $2p$, respectively, provided that all intermediate quantities are calculated with sufficient accuracy. It follows that to implement an FEM of the order p, the corresponding integrals must be computed with an accuracy of at least the order $p + 1$. The most economical way to calculate such integrals rests on the use of quadrature rules of the Gaussian type. In the one-dimensional case, these quadrature rules are known analytically. Using their Cartesian product, analytical quadrature rules can be constructed for rectangular hyperparallelepipeds, however, they are not optimal. Such quadrature rules, characterized by almost minimal numbers of nodes, are reported in [41]. Quadrature rules are also known for curvilinear domains [42, 43]. In the lower order cases, expressed in terms of radicals, the weights and nodes of quadrature rules are exact. Such solutions for quadrature rules up to the fifth-order on a triangle and up to the third order on a tetrahedron are reported in [44], while quadrature rules of the second- and third-orders on an arbitrary simplex are reported in [45]. The use of Gröbner bases makes it possible to derive general solutions to systems of nonlinear algebraic equations for quadrature rules up to the fifth order on a simplex (for example, using the PolynomialSystem program implemented in the MAPLE system). This approach, however, cannot provide solutions beyond the eighth order. It is noteworthy that for multidimensional integrals, various quadrature rules of the Grundmann-Möller and Newton-Cotes types are known (see, e.g., Refs. [46, 47]). Gaussian quadrature rules are reported in many papers (for example, [48–66]).

3.2 Setting of the Problem

Consider a self-adjoint BVP for the d dimensional elliptic differential equation of the second-order:

$$(D - E)\, \Phi(z) \equiv \left(-\frac{1}{g_0(z)} \sum_{ij=1}^{d} \frac{\partial}{\partial z_i} g_{ij}(z) \frac{\partial}{\partial z_j} + V(z) - E \right) \Phi(z) = 0. \qquad (3.1)$$

For the principal part coefficients of Eq. (3.1) the condition of uniform ellipticity holds in the bounded domain $z = (z_1, \ldots, z_d) \in \Omega$ of the Euclidean space $\mathcal{R}^d$, i.e. the constants $\mu > 0$, $\nu > 0$ exist such that

$$\mu \xi^2 \leq \sum_{ij=1}^{d} g_{ij}(z) \xi_i \xi_j \leq \nu \xi^2, \quad \xi^2 = \sum_{i=1}^{d} \xi_i^2, \quad \forall \xi \in \mathcal{R}^d. \qquad (3.2)$$

The left-hand side of this inequality expresses the requirement of ellipticity, while the right-hand side expresses the boundedness of the coefficients $g_{ij}(z)$. It is also assumed that $g_0(z) > 0$, $g_{ji}(z) = g_{ij}(z)$ and $V(z)$ are real-valued functions, continuous together with their generalized derivatives to a given order in the domain $z \in \bar{\Omega} = \Omega \cup \partial\Omega$ with the piecewise continuous boundary $S = \partial\Omega$, which provide the existence of nontrivial solutions obeying the boundary conditions [31, 67] of the first kind

$$\Phi(z)|_S = 0, \qquad (3.3)$$

or the second kind

$$\frac{\partial \Phi(z)}{\partial n_D}\bigg|_S = 0, \quad \frac{\partial \Phi(z)}{\partial n_D} = \sum_{ij=1}^{d} (\hat{n}, \hat{e}_i) g_{ij}(z) \frac{\partial \Phi(z)}{\partial z_j}, \qquad (3.4)$$

where $\frac{\partial \Phi_m(z)}{\partial n_D}$ is the derivative along the conormal direction, $\hat{n}$ is the outer normal to the boundary of the domain $S = \partial\Omega$, $\hat{e}_i$ is the unit vector of $z = \sum_{i=1}^{d} \hat{e}_i z_i$, $(\hat{n}, \hat{e}_i)$ is the scalar product in $\mathcal{R}^d$ $((\hat{n}, \hat{e}_1) = 1$ for $d = 1)$.

For a discrete spectrum problem the functions $\Phi_m(z)$ from the Sobolev space $H_2^{s \geq 1}(\Omega)$, $\Phi_m(z) \in H_2^{s \geq 1}(\Omega)$, corresponding to the real eigenvalues E: $E_1 \leq E_2 \leq \cdots \leq E_m \leq \cdots$ satisfy the conditions of normalization and orthogonality

$$\langle \Phi_m(z) | \Phi_{m'}(z) \rangle = \int_\Omega g_0(z) \Phi_m(z) \Phi_{m'}(z) dz = \delta_{mm'}, \quad dz = dz_1 \cdots dz_d. \qquad (3.5)$$

The FEM solution of the BVPs (3.1)–(3.5) is reduced to the determination of stationary points of the variational functional [10, 31]

$$\Xi(\Phi_m, E_m, z) \equiv \int_{\Omega} g_0(z)\Phi_m(z)\,(D - E_m)]\Phi_m(z)dz = \Pi(\Phi_m, E_m, z), \qquad (3.6)$$

where $\Pi(\Phi_m, E_m, z)$ is the symmetric quadratic functional

$$\Pi(\Phi_m, E_m, z) = \int_{\Omega}\left[\sum_{ij=1}^{d} g_{ij}(z)\frac{\partial \Phi_m(z)}{\partial z_i}\frac{\partial \Phi_m(z)}{\partial z_j}\right.$$

$$\left. + g_0(z)\Phi_m(z)(V(z) - E_m)\Phi_m(z)\right]dz. \qquad (3.7)$$

In the FEM the polyhedral domain $\Omega = \bigcup_{q=1}^{Q} \Delta_q$ is covered with the subdomains Δ_q (usually d-simplexes or d-parallelepipeds) called finite elements, and on each of them the shape functions $\varphi_{lq}^{p'}(z)$ are constructed using multivariable Hermitian interpolation polynomials and their partial derivatives and their derivatives along the normals to the boundaries of the finite elements. The piecewise polynomial functions $N_l^{p'}(z)$ in the domain Ω are constructed by joining the shape functions $\varphi_{lq}^{p'}(z)$ in the finite element Δ_q:

$$N_l^{p'}(z) = \{\varphi_{lq}^{p'}(z),\, A_l \in \Delta_q;\, 0,\, A_l \notin \Delta_q\}, \qquad (3.8)$$

and possess the following properties: functions $N_l^{p'}(z)$ are continuous in the domain Ω; the functions $N_l^{p'}(z)$ or their partial derivatives equal 1 in one of the points A_l and zero in the rest points. The detailed description is done in Sect. 3.3.

The functions $N_l^{p'}(z)$ form a basis in the space of polynomials of the p'th order. Now, the function $\Phi(z) \in \mathcal{H}^{\kappa'_{d-1}+1\geq 1}(\Omega)$, where κ'_{d-1} is a maximal order of continuous derivatives of $N_l^{p'}(z)$ in face of dimension $d-1$ of d-dimensional finite element, is approximated by a finite sum of piecewise basis functions $N_l^{p'}(z)$

$$\Phi^h(z) = \sum_{l=1}^{L} \Phi_l^h N_l^{p'}(z). \qquad (3.9)$$

After substituting the expansion (3.9) into the variational functional (3.6) and minimizing it [8, 10], we obtain the generalized eigenvalue problem

$$\mathbf{A}^{p'}\mathbf{\Phi}^h = \varepsilon^h \mathbf{B}^{p'}\mathbf{\Phi}^h. \qquad (3.10)$$

Here $\mathbf{A}^{p'}$ is the symmetric stiffness matrix, $\mathbf{B}^{p'}$ is the positive definite symmetric mass matrix, and ε^h is the corresponding eigenvalue. The matrices $\mathbf{A}^{p'}$ and $\mathbf{B}^{p'}$ have the form:

$$\mathbf{A}^{p'} = \{a_{ll'}^{p'}\}_{ll'=1}^{L}, \quad \mathbf{B}^{p'} = \{b_{ll'}^{p'}\}_{ll'=1}^{L}, \qquad (3.11)$$

where the matrix elements $a_{ll'}^{p'}$ and $b_{ll'}^{p'}$ are have the form

$$a_{ll'}^{p'} = \int_\Omega N_l^{p'}(z) N_{l'}^{p'}(z) U(z) g_0(z) dz + \sum_{i,j=1}^d \int_\Omega \frac{\partial N_l^{p'}(z)}{\partial z_i} \frac{\partial N_{l'}^{p'}(z)}{\partial z_j} g_{ij}(z) dz$$

$$= \sum_{q=1}^Q \left(\int_{\Delta_q} \varphi_{lq}^{p'}(z) \varphi_{l'q}^{p'}(z) U(z) g_0(z) dz \right.$$

$$\left. + \sum_{i,j=1}^d \int_{\Delta_q} \frac{\partial \varphi_{lq}^{p'}(z)}{\partial z_i} \frac{\partial \varphi_{l'q}^{p'}(z)}{\partial z_j} g_{ij}(z) dz \right),$$

$$b_{ll'}^{p'} = \int_\Omega N_l^{p'}(z) N_{l'}^{p'}(z) g_0(z) dz = \sum_{q=1}^Q \int_{\Delta_q} \varphi_{lq}^{p'}(z) \varphi_{l'q}^{p'}(z) g_0(z) dz. \quad (3.12)$$

The integration in (3.12) is performed over some subdomains Δ_q, where the basis functions under integral $N_l^{p'}(z)$ composed by matching (3.8) of shape functions $\varphi_{lq}^{p'}(z)$ are not equal zero. One of most effective numerical method is Gauss type quadrature rules, which for the d-simplex are presented in Appendix B.

The deviation of the approximate solution E_m^h, $\Phi_m^h(z) \in \mathcal{H}^{\kappa'_{d-1}+1 \geq 1}(\Omega_h)$ from the exact one E_m, $\Phi_m(z) \in \mathcal{H}^2(\Omega)$ is theoretically estimated as [8, 31]

$$\left| E_m - E_m^h \right| \leq c_1 h^{2p'}, \quad \left\| \Phi_m(z) - \Phi_m^h(z) \right\|_0 \leq c_2 h^{p'+1}, \quad (3.13)$$

where $\|\Phi_m(z)\|_0^2 = \int_\Omega g_0(z) \overline{\Phi}_m(z) \Phi_m(z) dz$, h is the maximal size of the finite element Δ_q, p' is the order of the FEM, m is the number of the eigenvalue, c_1 and c_2 are coefficients independent of h.

3.3 The Shape Functions

3.3.1 One Dimensional Lagrange and Hermite Interpolation Polynomials

In each element Δ_q we define the sub-grid [39]

$$\Omega_q^{h_q(z)}[z_q^{\min}, z_q^{\max}] = \{z_{0q} = z_q^{\min}, z_{rq}, r = 1, \ldots, p-1, z_{pq} = z_q^{\max}\} \quad (3.14)$$

with the nodal points z_{rq}.

As a set of shape functions $\{N_{lq}^{p'}(z)\}_{l=0}^{l^{\max}}$, where $l^{\max} + 1$ is number of shape functions, we will use the Hermite interpolation polynomials (HIPs) $\{\{\varphi_{rq}^{\kappa p'}(z)\}_{\kappa=0}^{\kappa_r^{\max}-1}\}_{r=0}^p$ in the nodes z_{rq}, $r = 0, \ldots, p$ of the grid (3.14). The values of the functions $\varphi_{rq}^{\kappa p'}(z)$ with their derivatives up to the order $(\kappa_r^{\max} - 1)$, i.e. $\kappa = 0, \ldots, \kappa_r^{\max} - 1$, where $\kappa_r^{\max}$ is referred to as the multiplicity of the node z_r, are determined by the expressions [25]

$$\varphi_{rq}^{\kappa p'}(z_{r'}) = \delta_{rr'}\delta_{\kappa 0}, \qquad \left.\frac{d^{\kappa'}\varphi_{rq}^{\kappa p'}(z)}{dz^{\kappa'}}\right|_{z=z_{r'}} = \delta_{rr'}\delta_{\kappa\kappa'}. \tag{3.15}$$

In particular case, the shape functions called by Lagrange interpolation polynomials (LIPs) are determined only their values on subgrid $\Omega_q^{h_q(z)}$ and have the simple form

$$\varphi_{rq}^{\kappa p'}(z_{r'q}) = \delta_{rr'}, \qquad \varphi_{rq}^{\kappa p'}(z) = \prod_{r'=0,r'\neq r}^{p} \left(\frac{z - z_{r'q}}{z_{rq} - z_{r'q}}\right) \tag{3.16}$$

with $\kappa = 0$, $p' = p$.

To calculate the HIPs the auxiliary weight function

$$w_{rq}(z) = \prod_{r'=0,r'\neq r}^{p} \left(\frac{z - z_{r'q}}{z_{rq} - z_{r'q}}\right)^{\kappa_{r'}^{\max}}, \qquad w_{rq}(z_{rq}) = 1 \tag{3.17}$$

is used. The weight function derivatives can be presented as a product

$$\frac{d^{\kappa} w_{rq}(z)}{dz^{\kappa}} = w_{rq}(z)g_{rq}^{\kappa}(z), \tag{3.18}$$

where the factor $g_{rq}^{\kappa}(z)$ is calculated by means of the recurrence relations

$$g_{rq}^{\kappa}(z) = \frac{dg_{rq}^{\kappa-1}(z)}{dz} + g_{rq}^{1}(z)g_{rq}^{\kappa-1}(z), \tag{3.19}$$

with the initial conditions

$$g_{rq}^{0}(z) = 1, \quad g_{rq}^{1}(z) \equiv \frac{1}{w_{rq}(z)}\frac{dw_{rq}(z)}{dz} = \sum_{r'=0,r'\neq r}^{p} \frac{\kappa_{r'q}^{\max}}{z - z_{r'q}}. \tag{3.20}$$

We will seek for the HIPs $\varphi_{rq}^{\kappa}(z)$ in the following form:

$$\varphi_{rq}^{\kappa p'}(z) = w_{rq}(z) \sum_{\kappa'=0}^{\kappa_r^{\max}-1} a_{rq}^{\kappa,\kappa'}(z - z_{rq})^{\kappa'}. \tag{3.21}$$

Differentiating the function (3.21) by z at the point of z_{rq} and using Eq. (3.17), we obtain

$$\left.\frac{d^{\kappa'}\varphi_{rq}^{\kappa p'}(z)}{dz^{\kappa'}}\right|_{z=z_{rq}} = \sum_{\kappa''=0}^{\kappa'} \frac{\kappa'!}{\kappa''!(\kappa'-\kappa'')!} g_{rq}^{\kappa'-\kappa''}(z_{rq})a_{rq}^{\kappa,\kappa''}\kappa''!. \tag{3.22}$$

Hence we arrive at the expression for the coefficients $a_r^{\kappa,\kappa'}$

$$a_{rq}^{\kappa,\kappa'} = \left(\frac{d^{\kappa'} \varphi_{rq}^{\kappa p'}(z)}{dz^{\kappa'}} \bigg|_{z=z_{rq}} - \sum_{\kappa''=0}^{\kappa'-1} \frac{\kappa'!}{(\kappa'-\kappa'')!} g_{rq}^{\kappa'-\kappa''}(z_{rq}) a_{rq}^{\kappa,\kappa''} \right) / \kappa'. \qquad (3.23)$$

Taking Eq. (3.15) into account, we finally get:

$$a_{rq}^{\kappa,\kappa'} = \begin{cases} 0, & \kappa' < \kappa, \\ 1/\kappa'!, & \kappa' = \kappa, \\ -\sum_{\kappa''=\kappa}^{\kappa'-1} \frac{1}{(\kappa'-\kappa'')!} g_{rq}^{\kappa'-\kappa''}(z_{rq}) a_{rq}^{\kappa,\kappa''}, & \kappa' > \kappa. \end{cases} \qquad (3.24)$$

Note that all degrees of HIPs $\varphi_{rq}^{\kappa p'}(z)$ do not depend on κ and equal $p' = \sum_{r'=0}^{p} \kappa_r^{\max} - 1$ and form a basis in the space of polynomials having the degree p'.

The HIPs with similar multiplicity $\kappa_r^{\max} = \kappa^{\max}$ also form a basis

$$N_{\kappa^{\max}r+\kappa}(z, z_q^{\min}, z_q^{\max}) = \varphi_{rq}^{\kappa p'}(z), \quad r = 0, \dots, p, \quad \kappa = 0, \dots, \kappa^{\max} - 1 \quad (3.25)$$

in the space of polynomials having the degree $p' = \kappa^{\max}(p+1) - 1$ in the element $z \in [z_q^{\min}, z_q^{\max}]$ that have continuous derivatives up to the order $\kappa^{\max} - 1$ at the boundary points $z_q^{\min}$ and $z_q^{\max}$ of the element $z \in [z_q^{\min}, z_q^{\max}]$. The HIPs at $\kappa^{\max} = 1, 2, 3$ and $p = 4$ are shown in Fig. 3.1. It is seen that the values of HIP $N_{\kappa^{\max}p+\kappa}(z, z_q^{\min}, z_q^{\max})$ and $N_\kappa(z, z_{q+1}^{\min}, z_{q+1}^{\max})$ (at $r = p$ and $r = 0$) and their derivatives up to the order $\kappa^{\max} - 1$ coincide at the mutual point $z_q^{\max} = z_{q+1}^{\min}$ of the adjacent elements. Moreover, the boundary

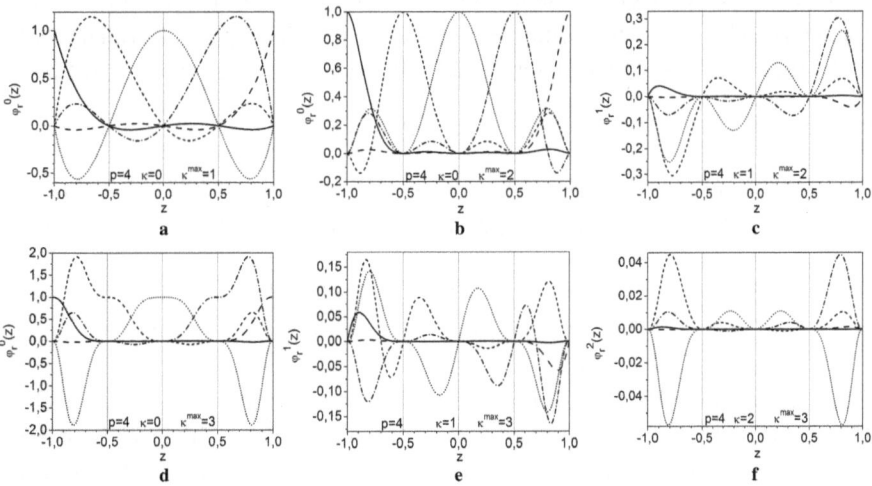

Fig. 3.1 The HIP coinciding at $\kappa^{\max} = 1$ with the LIP (**a**) and HIPs at $\kappa^{\max} = 2$ (**b, c**) and $\kappa^{\max} = 3$ (**d, e, f**). Here $p + 1 = 5$ is the number of nodes in the subinterval, $\Delta_j = [z_q^{\min} = -1, z_q^{\max} = 1]$. The grid nodes z_{rq} are shown by vertical lines

points are nodes (zeros) of multiplicity $\kappa^{\max}$ of other HIPs, irrespective of the length of elements of $[z_q^{\min}, z_q^{\max}]$ and $[z_{q+1}^{\min}, z_{q+1}^{\max}]$. This allows construction of a basis of piecewise and polynomial functions having continuous derivatives to the order of $\kappa^{\max} - 1$ in any set $\Omega = \bigcup_{q=1}^{Q} \Delta_q$ of elements $\Delta_q = [z_q^{\min}, z_q^{\max} \equiv z_{q+1}^{\min}]$.

We consider a discrete representation of the solutions $\Phi(z)$ of the problem (3.1), (3.3), (3.5) at $d = 1$ reduced by means of the FEM to the variational functional (3.6), (3.7) on the finite-element grid,

$$\Omega^{h_q(z)}[z^{\min}, z^{\max}] = [z_0 = z^{\min}, z_l, l = 1, \dots, np - 1, z_{np} = z^{\max}], \quad (3.26)$$

with the mesh points $z_l = z_{qp} = z_q^{\max} \equiv z_{q+1}^{\min}$ of the grid $\Omega^{h_q(z)}[z^{\min}, z^{\max}]$ and the nodal points $z_l = z_{(q-1)p+r}, r = 0, \dots, p$ of the sub-grids $\Omega_q^{h_q(z)}[z_q^{\min}, z_q^{\max}], q = 1, \dots, n$, determined by Eq. (3.14).

The solutions $\Phi^h(z) \approx \Phi(z)$ are sought for in the form of a finite sum over the basis of local functions $N_\mu^{p'}(z)$ at each nodal point $z = z_l$ of the grid $\Omega^{h_q(z)}[z^{\min}, z^{\max}]$:

$$\Phi^h(z) = \sum_{\mu=0}^{L-1} \Phi_\mu^h N_\mu^{p'}(z), \quad \Phi^h(z_l) = \Phi_{l\kappa^{\max}}^h, \quad \frac{d^\kappa \Phi^h(z)}{dz^\kappa}\bigg|_{z=z_l} = \Phi_{l\kappa^{\max}+\kappa}^h, \quad (3.27)$$

where $L = (pn + 1)\kappa^{\max}$ is the number of local functions and Φ_μ^h at $\mu = l\kappa^{\max} + \kappa$ are the nodal values of the κth derivatives of the function $\Phi^h(z)$ (including the function $\Phi^h(z)$ itself for $\kappa = 0$) at the points z_l.

The local functions $N_\mu^{p'}(z) \equiv N_{l\kappa^{\max}+\kappa}^{p'}(z)$ are piecewise polynomials of the given order p', their derivative of the order κ at the node z_l equals one, and the derivative of the order $\kappa' \neq \kappa$ at this node equals zero, while the values of the function $N_\mu^{p'}(z)$ with all its derivatives up to the order $(\kappa^{\max} - 1)$ equal zero at all other nodes $z_{l'} \neq z_l$ of the grid $\Omega^{h_q(z)}$, i.e.,

$$\frac{d^\kappa N_{l'\kappa^{\max}+\kappa'}^{p'}(z)}{dz^\kappa}\bigg|_{z=z_l} = \delta_{ll'}\delta_{\kappa\kappa'}, l = 0, \dots, np, \kappa = 0, \dots, \kappa^{\max} - 1.$$

For the nodes z_l of the grid that do not coincide with the mesh points $z_q^{\max}$, i.e., at $l \neq qp$, $q = 1 \dots n - 1$, the polynomial $N_\mu^{p'}(z)$ at $\mu = ((q-1)p + r)\kappa^{\max} + \kappa$ has the form

$$N_{(p(q-1)+r)\kappa^{\max}+\kappa}^{p'}(z) = \begin{cases} N_{\kappa^{\max}r+\kappa}^{p'}(z, z_q^{\min}, z_q^{\max}), & z \in \Delta_q, \\ 0, & z \notin \Delta_q, \end{cases} \quad (3.28)$$

i.e., it is defined as the HIP $N_{\kappa^{\max}r+\kappa}^{p'}(z, z_q^{\min}, z_q^{\max})$ in the interval $z \in \Delta_q$ and zero otherwise. Since the points $z_q^{\min}$ and $z_q^{\max}$ are nodes of multiplicity $\kappa^{\max}$, such piecewise polynomial functions and their derivatives up to the order $\kappa^{\max} - 1$ are continuous in the entire interval Ω. In Fig. 3.1 such HIPs are plotted by dotted, short-dashed and dot-dashed lines.

3.3.2 Lagrange Interpolation Polynomials on Simplex

In FEM the domain $\Omega = \Omega_h(z) = \bigcup_{q=1}^{Q} \Delta_q$, specified as a polyhedral domain, is covered with finite elements, in the present case the simplexes Δ_q with $d+1$ vertices $\hat{z}_i = (\hat{z}_{i1}, \hat{z}_{i2}, \ldots, \hat{z}_{id})$ with $i = 0, \ldots, d$. Each edge of the simplex Δ_q is divided into p equal parts and the families of parallel hyperplanes $H(i, k)$ are drawn, numbered with the integers $k = 0, \ldots, p$, starting from the corresponding face, e.g., as shown for $d = 2$ in Fig. 3.2 (see also [31]). The equation of the hyperplane $H(i, k)$ is: $H(i; z) - k/p = 0$, where $H(i; z)$ is a linear function of z.

The node points of hyperplanes crossing A_r are enumerated with sets of integers $[n_0, \ldots, n_d]$, $n_i \geq 0$, $n_0 + \cdots + n_d = p$, where n_i, $i = 0, 1, \ldots, d$ are the numbers of hyperplanes, parallel to the simplex face, not containing the ith vertex $\hat{z}_i = (\hat{z}_{i1}, \ldots, \hat{z}_{id})$. The coordinates $\xi_r = (\xi_{r1}, \ldots, \xi_{rd})$ of the node point $A_r \in \Delta_q$ are calculated using the formula

$$(\xi_{r1}, \ldots, \xi_{rd}) = \sum_{j=0}^{d} (\hat{z}_{j1}, \ldots, \hat{z}_{jd}) \frac{n_j}{p} \qquad (3.29)$$

from the coordinates of the vertices $\hat{z}_j = (\hat{z}_{j1}, \ldots, \hat{z}_{jd})$. Then the LIP $\varphi_{rq}^{\kappa p'}(z)$ of order $p' = p$ and $\kappa = 0$ equal to one at the point A_r with the coordinates $\xi_r = (\xi_{r1}, \ldots, \xi_{rd})$, characterized by the numbers $[n_0, n_1, \ldots, n_d]$, and equal to zero at the remaining points

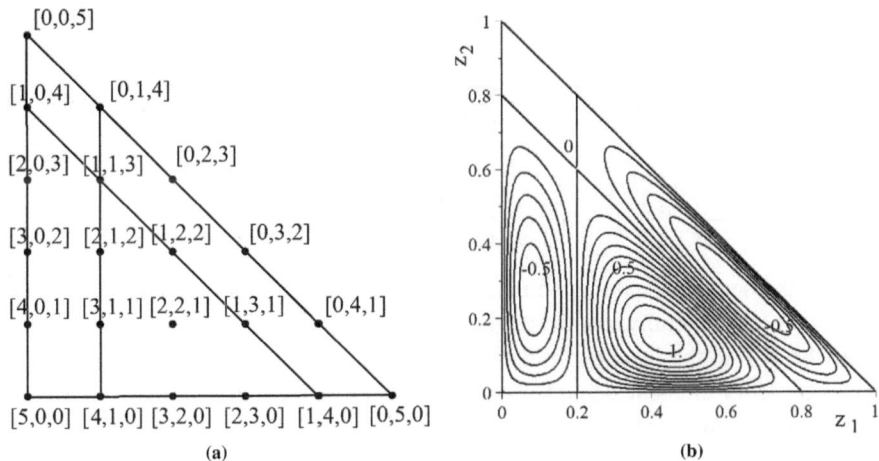

(a) (b)

Fig. 3.2 a Enumeration of nodes $A_r, r = 1, \ldots, (p+1)(p+2)/2$ with sets of numbers $[n_0, n_1, n_2]$ for the standard (canonical) triangle element Δ in the method with the fifth-order LIP $p' = p = 5$ at $d = 2$. The lines (five crossing straight lines) are zeros of LIP $\varphi_{14}(z')$ from (3.35), equal to 1 at the point labeled with the number triple $[n_0, n_1, n_2] = [2, 2, 1]$. **b** LIP isolines of $\varphi_{14}(z')$

$\xi_{r'}$, i.e., $\varphi_{rq}^{\kappa p'}(\xi_{r'}) = \delta_{rr'}$, has the form

$$\varphi_{rq}^{\kappa p'}(z) = \prod_{i=0}^{d} \prod_{n_i'=0}^{n_i-1} \frac{H(i;z) - n_i'/p}{H(i;\xi_r) - n_i'/p}. \tag{3.30}$$

3.3.3 The Economical Implementation of Finite Element Method

Note that the construction of the HIP $\varphi_{rq}^{\kappa p'}(z)$, where $\kappa \equiv \kappa_1, \ldots, \kappa_d$, with the fixed values of the functions $\{\varphi_{rq}^{\kappa p'}(\xi_{r'})\}$ and the derivatives $\{\partial_{\bullet} \varphi_{rq}^{\kappa p'}(z)|_{z=\xi_{r'}}\}$ at the nodes $\xi_{r'}$, already at $d = 2$ leads to cumbersome expressions, improper for FEM using nonuniform mesh.

The economical implementation of FEM is the following:

1. The calculations are performed in the local (reference) coordinates z', in which the coordinates of the simplex vertices are the following: $\hat{z}_j' = (\hat{z}_{j1}', \ldots, \hat{z}_{jd}')$, $\hat{z}_{jk}' = \delta_{jk}$,

2. The HIP in the physical coordinates z in the mesh is sought in the form of linear combinations of polynomials in the local coordinates z', the transition to the physical coordinates executed only at the stage of numerical solution of a particular BVP (3.1)–(3.6),

3. The calculation of FEM integrals is executed in the local coordinates.

Let us construct the HIP on an arbitrary d-dimensional simplex Δ_q with the $d+1$ vertices $\hat{z}_i = (\hat{z}_{i1}, \hat{z}_{i2}, \ldots, \hat{z}_{id})$, $i = 0, \ldots, d$. For this purpose we introduce the local coordinate system $z' = (z_1', z_2', \ldots, z_d') \in \mathcal{R}^d$, in which the coordinates of the simplex vertices are the following: $\hat{z}_i' = (\hat{z}_{ik}' = \delta_{ik}, k = 1, \ldots, d)$. The direct and inverse relations between coordinates and differentiation operators are given by the formulas:

$$z_i = \hat{z}_{0i} + \sum_{j=1}^{d} \hat{J}_{ij} z_j', \quad z_i' = \sum_{j=1}^{d} (J^{-1})_{ij}(z_j - \hat{z}_{0j}), \quad \hat{J}_{ij} = \hat{z}_{ji} - \hat{z}_{0i}, \tag{3.31}$$

$$\frac{\partial}{\partial z_i'} = \sum_{j=1}^{d} \hat{J}_{ji} \frac{\partial}{\partial z_j}, \quad \frac{\partial}{\partial z_i} = \sum_{j=1}^{d} (\hat{J}^{-1})_{ji} \frac{\partial}{\partial z_j'}, \quad i = 1, \ldots, d. \tag{3.32}$$

In this case the derivatives along the normal to the element boundary in the physical coordinate system are, generally, not those in the local coordinates z' (see Fig. 3.3). When constructing the HIP in the local coordinates z' one has to recalculate the fixed derivatives at the nodes $\xi_{r'}$ of the element Δ_q to the nodes $\xi_{r'}'$ of the element Δ, using the matrices $\hat{J}^{-1}$, given by cumbersome expressions.

The integrals that enter the variational functional (3.6) on the domain $\Omega_h(z) = \bigcup_{q=1}^{Q} \Delta_q$, are expressed via the integrals, calculated on the element Δ_q, and recalculated to the local coordinates z' on the element Δ,

$$\int_{\Delta_q} g_0(z) \varphi_{rq}^{\kappa p'}(z) \varphi_{r'q}^{\kappa'' p'}(z) U(z) dz$$

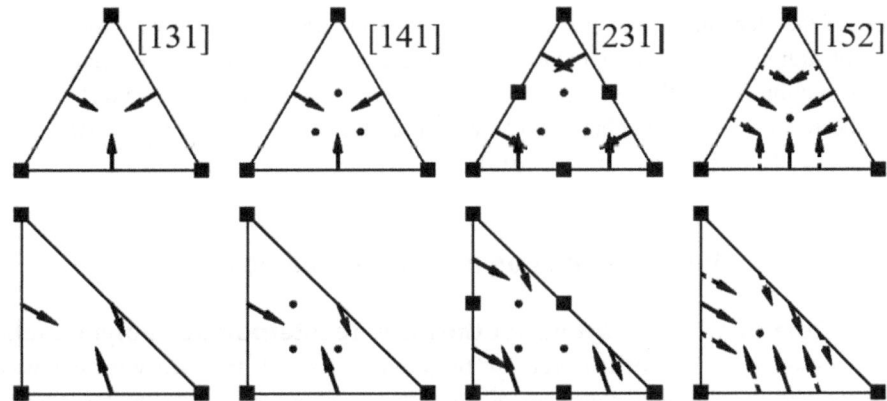

Fig. 3.3 Schematic diagram of the conditions on the element Δ_q (upper panel) and Δ (lower panel) for constructing the basis of two dimensional HIPs $[p\kappa_{max}\kappa_1']$: [131], [141], [231], [152] of order $p' = 5, 7, 8, 9$, respectively. The squares are the points ξ_r', where the values of the functions and their derivatives are fixed according to the conditions (3.36); the solid (dashed) arrows begin at the points η_s', where the values of the first (second) derivative in the direction of the normal in the physical coordinates are fixed, according to the condition (3.37), respectively; the circles are the points ζ_s', where the values of the functions are fixed according to the condition (3.38)

$$= J \int_\Delta g_0(z(z'))\varphi_{rq}^{\kappa p'}(z')\varphi_{r'q}^{\kappa'' p'}(z')U(z(z'))dz', \qquad (3.33)$$

$$\int_{\Delta_q} g_{s_1 s_2}(z)\frac{\partial \varphi_{rq}^{\kappa p'}(z)}{\partial z_{s_1}} \frac{\partial \varphi_{r'q}^{\kappa'' p'}(z)}{\partial z_{s_2}}dz$$

$$= J \sum_{t_1,t_2=1}^{d} \hat{J}_{s_1 s_2;t_1 t_2}^{-1} \int_\Delta g_{s_1 s_2}(z(z'))\frac{\partial \varphi_{rq}^{\kappa p'}(z')}{\partial z_{t_1}'} \frac{\partial \varphi_{r'q}^{\kappa'' p'}(z')}{\partial z_{t_2}'}dz', \qquad (3.34)$$

where $J = \det \hat{J} > 0$ is the determinant of the matrix $\hat{J}$ from Eq. (3.31), $\hat{J}_{s_1 s_2;t_1 t_2}^{-1} = (\hat{J}^{-1})_{t_1 s_1}(\hat{J}^{-1})_{t_2 s_2}$, $dz' = dz_1' \cdots dz_d'$.

In the local coordinates the LIP $\varphi_{rq}^{\kappa p'}(z')$, is equal to one at the node point ξ_r', characterized by the numbers $[n_0, n_1, \ldots, n_d]$, and zero at the remaining node points $\xi_{r'}'$, i.e., $\varphi_{rq}^{\kappa p'}(\xi_{r'}') = \delta_{rr'}$, are determined by Eq. (3.30) at $H(0; z') = 1 - z_1' - \cdots - z_d'$, $H(i; z') = z_i', i = 1, \ldots, d$:

$$\varphi_{rq}^{\kappa p'}(z') = \left(\prod_{i=1}^{d}\prod_{n_i'=0}^{n_i-1} \frac{z_i' - n_i'/p}{n_i/p - n_i'/p}\right)\left(\prod_{n_0'=0}^{n_0-1} \frac{1 - z_1' - \cdots - z_d' - n_0'/p}{n_0/p - n_0'/p}\right). \qquad (3.35)$$

Setting the numerators in Eq. (3.35) equal to zero, yields the families of equations for the straight lines, directed "horizontally", "vertically", and "diagonally" in the local coordinate system of the element Δ, which is related by the affine transformation with the "oblique" family of straight lines of the element Δ_q. In Fig. 3.2 an example is presented that illustrates the construction of the LIP at $d = 2, r, r' = 1, \ldots, (p+1)(p+2)/2, p = 5$ on the element Δ in the form of a rectangular triangle with the vertices $\hat{z}_0' = (\hat{z}_{01}', \hat{z}_{02}') = (0, 0)$, $\hat{z}_1' = (\hat{z}_{11}', \hat{z}_{12}') = (1, 0)$, $\hat{z}_2' = (\hat{z}_{21}', \hat{z}_{22}') = (0, 1)$.

3.3.4   Hermite Interpolation Polynomials on Simplex

3.3.4.1 Determining Conditions for the Hermite Interpolation Polynomials

Let us construct the HIP of the order p' by joining of which the piecewise polynomial functions with the continuous derivatives up to the given order κ_{d-1}' on the boundaries of the finite elements can be obtained.

Following the above (in a way similar to that for LIPs), the values of the functions themselves and their derivatives up to the order of $\kappa_{\max} - 1$ at the node points ξ_r' are specified by the conditions C1

$$\tilde{\varphi}_{rq}^{\kappa p'}(\xi_{r'}') = \delta_{rr'}\delta_{\kappa_1 0} \cdots \delta_{\kappa_d 0}, \quad \kappa = \kappa_1, \ldots, \kappa_d \tag{3.36}$$

$$\left. \frac{\partial^{|\mu|} \tilde{\varphi}_{rq}^{\kappa p'}(z')}{\partial z_1'^{\mu_1} \cdots \partial z_d'^{\mu_d}} \right|_{z'=\xi_{r'}'} = \delta_{rr'}\delta_{\kappa_1 \mu_1} \cdots \delta_{\kappa_d \mu_d},$$

$$0 \le \kappa_1 + \kappa_2 + \cdots + \kappa_d \le \kappa_{\max} - 1, \quad 0 \le |\mu| \equiv \mu_1 + \mu_2 + \cdots + \mu_d \le \kappa_{\max} - 1.$$

Here $r, r' = 1, \cdots, (p+d)!/(p!d!)$ labels the node points of crossing of hyperplanes $H(i, k)$ from Sect. 3.3.2, and number of the conditions C1 we denote by NC1.

For $d > 1$ and $\kappa_{\max} > 1$, the number $N_{\kappa_{\max}, p', d}$ of HIPs of the order p' and the multiplicity of nodes $\kappa_{\max}$ are smaller than the number $N_{1, p', d}$ of the polynomials that form the basis in the space of polynomials of the order p'.

For unique determination of the polynomial basis, let us introduce additional conditions. To provide the continuity of derivatives on boundaries (edges, faces and etc, but not on points, where conditions C1 were applied) of the d-simplex, we put additional conditions C2

$$\left. \frac{\partial^k \hat{\varphi}_{sq}^{\kappa p'}(z')}{\partial n_{i(s')}^k} \right|_{z'=\eta_{s'}'} = \delta_{ss'}, \quad s = 1, \ldots, \text{NC2} + \text{NC3}, \quad s' = 1, \ldots, \text{NC2}, \tag{3.37}$$

where NC2 and NC3 are numbers of conditions C2 and C3, $\eta_{s'}' = (\eta_{s'1}', \ldots, \eta_{s'd}')$ are the chosen points lying on the faces of various dimensionalities (from 1 to $d - 1$) of the d-dimensional simplex Δ and $\partial/\partial n_{i(s)}$ is the directional derivative along the vector n_i, normal (generally, not parallel) to the corresponding ith face of the d-dimensional simplex Δ_q at point $\eta_{s'}$.

Calculating the number of independent parameters required to provide the continuity of derivatives to the order κ on the simplex face of dimension d', we determine its maximal value $\kappa'_{d'}$. For the unique determination of the polynomials, additional conditions C3 are imposed:

$$\check{\phi}_{sq}^{\kappa p'}(\zeta'_{s''}) = \delta_{ss''}, \quad s = 1, \ldots, NC2 + NC3, \quad s'' = NC2 + 1, \ldots, NC2 + NC3, \quad (3.38)$$

where $\zeta'_{s''} = (\zeta'_{s''1}, \ldots, \zeta'_{s''d}) \in \Delta$ are the chosen points belonging to the simplex without the boundary.

The number of conditions $T = NC2 + NC3$, whether the points $\eta'_{s'}$ and $\zeta'_{s''}$ belong to the corresponding faces of dimension $d' = 1, \ldots, d - 1$ and d- itself simplex, and possible values of k, depending on the dimension of the face, are determined from criteria ensuring the continuity of the piecewise polynomial functions $N_{l'}^{p'}(z)$ from (3.9) and their derivatives of order $\kappa'_{d'}$ in a neighborhood of the faces dimensions $d' = 1, \ldots, d-1$, using the combinatorial algorithm, which counts the number of polynomials of different dimensions and order.

Note that the derivatives along the normal to the boundary of the element in the original coordinate system are generally not derivatives in the local coordinates x' (see Fig. 3.3). When constructing HIPs in local coordinates z', it is required to recalculate the fixed derivatives at the nodes $\xi_{r'}$ of the element Δ_q into the nodes $\xi'_{r'}$ of the element Δ,

$$\frac{\partial}{\partial n_i} = f_{i1}\frac{\partial}{\partial z'_1} + \cdots + f_{id}\frac{\partial}{\partial z'_d}, \quad i = 1, \ldots, T, \quad (3.39)$$

where $f_{ij} = f_{ij}(\hat{z}_0, \ldots, \hat{z}_d)$ are functions of the coordinates of the vertices of Δ_q simplex calculated using the formulas (3.31)–(3.32).

So, the basis of HIPs are constructed from conditions C1, C2 and C3 obtained by formulas (3.36), (3.37), and (3.38), where for each HIP all right-hand-side terms equal to zero except one of them is equal to 1.

The algorithm for selecting the conditions is based on the strategy of "how many conditions for polynomials are there and how many are needed" and will be illustrated in Sect. 3.3.4.2.

Algorithm for determining the number of conditions:
Step 1. The number $N_{1,p',d'}$ of polynomials of d' variables of order p' is determined by the formula

$$N_{1,p',d'} = (d' + p')!/(d'!p'!).$$

The number $N_{\kappa_{max},p',d'}$ of conditions C1 on each face of dimension d' is determined by the formula

$$N_{\kappa_{max},p',d'} = N_{1,p,d'}N_{1,\kappa_{max}-1,d'}.$$

The number of d'-dimensional faces for a d-dimensional simplex is determined by the formula

Table 3.1 The total number of conditions C2 and C3 on each face of dimension d' of a simplex of dimension $d > d'$

p	$\kappa_{\max}$	p'	$d' = 2$	$d' = 3$	$d' = 4$	$d' = 5$	$d' = 6$	$d' = 7$	$d' = 8$
1	2	3	1	0	0	0	0	0	0
1	3	5	3	4	1	0	0	0	0
1	4	7	6	16	15	6	1	0	0
1	5	9	10	40	65	56	28	8	1

$$G_{d',d} = (d+1)!/((d'+1)!(d-d')!).$$

Then for the number $K_{d'}$ of conditions C2 and C3 on a simplex of dimension d' we have

$$K_{d'} = N_{1,p',d'} - N_{\kappa_{\max},p',d'}, \; K_0 = K_1 = 0.$$

Next, successively, $d' = 1, 2, \ldots, d-1$ redefine $K_{d'}$ as the total number of conditions C2 and C3 on each d'-dimensional face of a $d > d'$-dimensional simplex or d'-dimensional simplex without taking into account of conditions C2 and C3 on its faces, we use recurrence relation

$$K_{d'} \to K_{d'} - \sum_{d''=2}^{d'-1} K_{d''} G_{d'',d'}.$$

Step 2. Next, we calculate the number of polynomials $K'_{d'\kappa'_{d'}}$ to ensure the continuity of derivatives of order $\kappa'_{d'}$ in the neighborhood of a face of dimension d', which satisfies to the inequality $K'_{d'\kappa'_{d'}} \leq K_{d'}$. For the first derivative, the number $K'_{d'\kappa'_{d'}}$ is equal to the number of polynomials of order $p' - 1$ without the number of conditions C1 with node multiplicity $\kappa_{\max} - 1$.

Step 3. The number of conditions C2 is determined by the formula

$$T = \sum_{d'=1}^{d-1} G_{d',d} K'_{d'\kappa'_{d'}}$$

and completes the execution of Algorithm.

3.3.4.2 Example

As an example, let us consider the construction of the basis of HIPs of the ninth-order.

HIPs on triangle

1. For example, for $p = 1$ and $\kappa_{\max} = 5$, we have polynomials of the order $p' = 9$. There are 55 such polynomials on the triangle, the number of conditions C1 is $3 \times 15 = 45$. Thus, there are $K = 10$ additional conditions C2 and C3 for determining the basis of HIPs.

The degree $p' = 9$ of the polynomial in the tangential variable t on the side of the triangle coincides with the degree of the polynomial of two variables, and it requires $p' + 1 = 10$ parameters to determine it uniquely. These will be the values of the polynomial and its derivatives up to the fourth-order.

2. The first-order derivative $k' = 1$ with respect to the variable normal to the boundary will be a polynomial of degree $p' - k' = 8$, and for its unambiguous determination $p' - k' + 1 = 9$ conditions are required. However, it is defined only by $(\kappa_{\max} - k')(p + 1) = 8$ conditions for mixed derivatives $\frac{\partial}{\partial n}, \frac{\partial^2}{\partial n \partial \tau}, \frac{\partial^3}{\partial n \partial \tau^2}, \frac{\partial^4}{\partial n \partial \tau^3}$ specified at two vertices. The missing parameter can be determined by specifying the derivative in a direction not parallel to the triangle boundary at one of the points on its side. We take the derivative along the inner normal, at the midpoint of the triangle side.

3. The second-order derivative $k' = 2$ with respect to the variable normal to the boundary is a polynomial of degree $p' - k' = 7$, and for its unambiguous determination $p' - k' + 1 = 8$ conditions are required. However, it is defined only by $(\kappa_{\max} - k')(p + 1) = 6$ conditions for derivatives $\frac{\partial^2}{\partial n^2}, \frac{\partial^3}{\partial n^2 \partial \tau}, \frac{\partial^4}{\partial n^2 \partial \tau^2}$ specified at two vertices of the triangle. Thus, 6 parameters are required to determine the second derivative unambiguously. We take the second derivative along the inner normal, at the points on the side of the triangle dividing it in a ratio of 2:1.

4. Thus, $10 - 3 - 6 = 1$ parameter remains, which will not be enough for constructing a basis of piecewise polynomial functions with a continuous third derivative, i.e., there is one C3 equal to one at the point $\zeta = (1/3, 1/3)$.

As a result, the basis of the ninth-order HIPs on the triangle is determined by 45 conditions C1, 3 and 6 conditions C2 for the first and second derivatives, respectively, and 1 condition C3, which are illustrated in Fig. 3.4.

Note that the basis set of these polynomials has been presented as benchmark for our algorithm in Table 5 of Ref. [40]. HIPs [120] of $p' = 3$ order and [131] of order $p' = 5$ coincide with known ones [29] and [28] respectively.

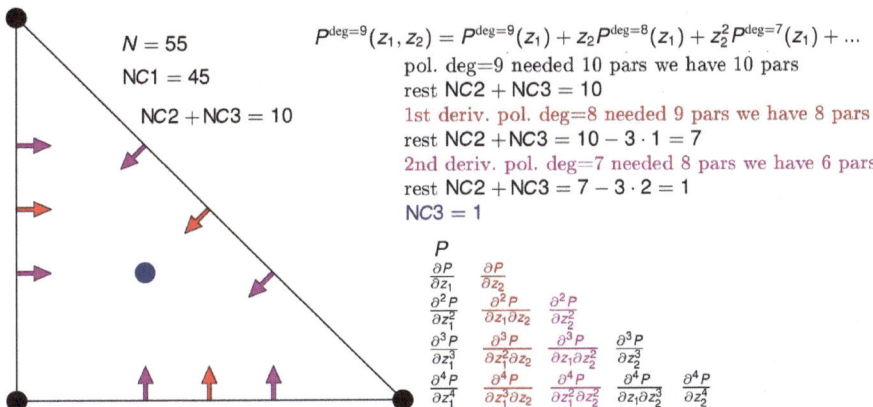

Fig. 3.4 The sets of left hand side of conditions C1, C2, and C3 for the ninth-order HIPs on a triangle

HIPs on a Tetrahedron

For the same example, for $p = 1$ and $\kappa_{\max} = 5$, we have polynomials of the order $p' = 9$. There are 220 such polynomials on a tetrahedron, and the number of conditions C1 is $4 \times 35 = 140$. In addition, on each edge, 2 conditions C2 for first derivatives in two directions, $2 \times 3 = 6$ conditions C2 for second directional derivatives (including the mixed one), and one condition C3 on each face are specified.

Thus, there are $K = 220 - 140 - 6 \times (2 + 6) - 4 \times 1 = 28$ additional conditions for determining conditions C2 on faces and conditions C3 inside the tetrahedron.

5. The first-order derivative $k' = 1$ with respect to the variable normal to the boundary will be a polynomial of degree $p' - k' = 8$, and 45 conditions are to determine it. However, it is determined by only 39 conditions: 10 mixed derivatives $\dfrac{\partial^{1+k_1+k_2}}{\partial n \partial \tau_1^{k_1} \partial \tau_2^{k_2}}$, $k_1, k_2 = 0, \ldots, 3$, $k_1 + k_2 \leq 3$ specified at three vertices of the tetrahedron, the derivatives $\dfrac{\partial}{\partial n}$, specified at the midpoints of the three edges of the tetrahedron, and mixed derivatives $\dfrac{\partial^2}{\partial n \partial \tau}$, which are specified at two points on each of the three edges of the tetrahedron. The missing 6 conditions C2 are determined by specifying the directional derivative not parallel to the tetrahedron face at six points on its face, but not on the edges.

6. Thus, $28 - 4 \times 6 = 4$ parameters remain, which will not be enough to construct a basis of piecewise polynomial functions with a continuous second derivative, i.e. there are four conditions C3 inside the tetrahedron.

The sets of left hand side of conditions (3.36), (3.37), and (3.38) called conditions C1, C2, and C3 for ninth-order HIPs on tetrahedron are illustrated in Fig. 3.5.

Note that these 3d HIPs are constructed in paper [68], where the following theorem is formulated:

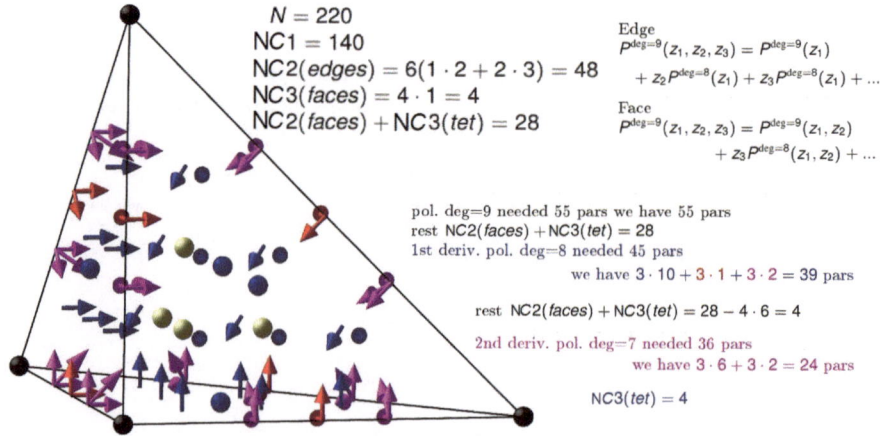

$N = 220$
$NC1 = 140$
$NC2(edges) = 6(1 \cdot 2 + 2 \cdot 3) = 48$
$NC3(faces) = 4 \cdot 1 = 4$
$NC2(faces) + NC3(tet) = 28$

Edge
$P^{\deg=9}(z_1, z_2, z_3) = P^{\deg=9}(z_1)$
    $+ z_2 P^{\deg=8}(z_1) + z_3 P^{\deg=8}(z_1) + \ldots$
Face
$P^{\deg=9}(z_1, z_2, z_3) = P^{\deg=9}(z_1, z_2)$
    $+ z_3 P^{\deg=8}(z_1, z_2) + \ldots$

pol. deg=9 needed 55 pars we have 55 pars
rest $NC2(faces) + NC3(tet) = 28$
1st deriv. pol. deg=8 needed 45 pars
    we have $3 \cdot 10 + 3 \cdot 1 + 3 \cdot 2 = 39$ pars

rest $NC2(faces) + NC3(tet) = 28 - 4 \cdot 6 = 4$

2nd deriv. pol. deg=7 needed 36 pars
    we have $3 \cdot 6 + 3 \cdot 2 = 24$ pars

$NC3(tet) = 4$

Fig. 3.5 The sets of left hand side of conditions C1, C2, and C3 for the ninth-order HIPs on a tetrahedron

Theorem 3.1 *The simplest polynomial on the d-dimensional simplex which generates piecewise polynomial and m-times continuously differentiable functions is of the degree* $n = 2^d m + 1$.

3.3.4.3 Algorithm for Determining the Hermite Interpolation Polynomials

The conditions C1, C2 and C3 obtained by formulas (3.36), (3.37), and (3.38) in the previous section are necessary but not sufficient for determining the set of HIPs, the piecewise polynomial functions composed of which preserve the continuity of derivatives up to a given order $\kappa'_{d'}$ due to the points $\eta'_{s'}$, $\zeta'_{s''}$ may be chosen incorrectly. In a general case the HIPs

$$\varphi_{tq}^{\kappa p'}(z_1, \ldots, z_d) = \{\tilde{\varphi}_{rq}^{\kappa p'}(z_1, \ldots, z_d), \hat{\varphi}_{s'q}^{\kappa p'}(z_1, \ldots, z_d), \check{\varphi}_{s''q}^{\kappa p'}(z_1, \ldots, z_d)\} \quad (3.40)$$

of the order p' on a d-simplex are sought in the form

$$\varphi_{tq}^{\kappa p'}(z_1, \ldots, z_d) = \sum_{i_1, \ldots, i_d = 0}^{i_1 + \cdots + i_d \leq p'} a_{i_1, \ldots, i_d}^{\kappa, p', t, q} z_1^{i_1} \times \cdots \times z_d^{i_d}. \quad (3.41)$$

Substituting expansion (3.41) into conditions C1, C2 and C3 obtained by formulas (3.36), (3.37), and (3.38), one obtain the system of linear algebraic equations

$$\mathbf{P}\mathbf{a}^{\kappa, p', t, q} = \mathbf{b}^{\kappa, p', t, q}, \quad (3.42)$$

where $\mathbf{a}^{\kappa, p', t, q}$ is a matrix composed of the expansion coefficients $a_{i_1, \ldots, i_d}^{\kappa, p', t, q}$ from (3.41), $\mathbf{P}$ is the matrix of coefficients at the components of the vector $\mathbf{a}^{\kappa, p', t, q}$, $\mathbf{b}^{\kappa, p', t, q}$ are the unit matrix of right-hand sides of the conditions (3.36), (3.37), and (3.38). The nondegeneracy of the matrix $\mathbf{P}$ is a sufficient condition for the existence of HIPs satisfying (3.36), (3.37), and (3.38).

The number of conditions obtained by the algorithm when the system of equations is compatible is presented in Tables 3.2 and 3.3. Explicit expressions for the corresponding HIPs are not presented here they are to be restored from the obtained conditions. The coordinates of the nodes for specifying conditions C2 (separately for each derivative of the order κ) and conditions C3 are chosen in the following way. We divide the edges into the required number of equal parts. For conditions C2 on the edge, these will be the necessary nodes, and for conditions C2 and C3 on the face or inside the tetrahedron, we draw parallel planes through these nodes, then the necessary nodes are their intersection points. If the number of nodes is not triangular or tetrahedral, then we exclude the corner nodes.

Table 3.2 The number of HIPs (N) on a triangle of order p' that provide the continuity of derivatives of the order κ'_1, numbers NC1, NC2 and NC3 of conditions C1, C2 and C3 versus κ_{max} and number NP2(s) of points on a side (s), where conditions C2 are applied

p	κ_{max}	κ'_1	p'	N	NC1	NP2(s)	NC2	NC3
1	1	0	1	3	3	0	0	0
1	2	0	3	10	9	0	0	1
1	3	1	5	21	18	1	3	0
1	4	1	7	36	30	1	3	3
2	3	1	8	45	36	2	6	3
1	5	2	9	55	45	1+2	9	1
2	4	1	11	78	60	2	6	12
1	6	2	11	78	63	1+2	9	6
1	7	3	13	105	84	1+2+3	18	3
1	8	3	15	136	108	1+2+3	18	10
1	9	4	17	171	135	1+2+3+4	30	6
1	$2k-1$	$k-1$	$4k-3$	$(2k-1)(4k-1)$	$3k(2k-1)$	$1+\cdots+$ $(k-1)$	$\frac{3}{2}k(k-1)$	$\frac{1}{2}(k-2)(k-1)$
1	$2k$	$k-1$	$4k-1$	$2k(4k+1)$	$3k(2k+1)$	$1+\cdots+$ $(k-1)$	$\frac{3}{2}k(k-1)$	$\frac{1}{2}k(k+1)$

Table 3.3 The number of HIPs (N) on a tetrahedron of order p' that provide the continuity of derivatives of the order κ'_1 and κ'_2 in vicinity of the edge and face and number NC1 of conditions C1, numbers NC2 of conditions C2 (on an edge (e), on a face (f) and total number) numbers NC3 of conditions C3 (on a face (f), inside a tetrahedron (t) and the total number) versus κ_{max}

κ_{max}	p'	κ'_1	κ'_2	N	NC1	NC2(e)	NC2(f)	NC2	NC3(f)	NC3(t)	NC3
1	1	0	0	4	4	0	0	0	0	0	0
2	3	0	0	20	16	0	0	0	1	0	4
3	5	1	0	56	40	2	0	12	0	4	4
4	7	1	0	120	80	2	0	12	3	16	28
5	9	2	1	220	140	$2+6=8$	6	72	1	4	8
6	11	2	1	364	224	$2+6=8$	12	96	6	20	44
7	13	3	1	560	336	$2+6+12=20$	10	160	3	52	64
8	15	3	1	816	480	$2+6+12=20$	18	192	10	104	144
9	17	4	2	1140	660	$2+6+12+20=40$	$15+25=40$	400	6	56	80

3.3.4.4 Hermite Interpolation Polynomials on a Simplex of an Arbitrary Dimension d

Here we present two examples obtained by generalizing these results to the case of an arbitrary dimension d. For these examples we choose $z_0 = 1 - z_1 - \cdots - z_d$, $i_k \neq i_l$, $i_k = 0, \ldots, d$.

1. Generalization of the basis of the third order HIPs $[p\kappa_{max}\kappa'_1 \cdots \kappa'_{d-1}] = [120 \cdots 0]$, $p = 1$, $\kappa_{max} = 2$, $p' = 3$, known as the Hermitian d-simplex [31].

To determine it we apply conditions C1 in the d-simplex vertices and one conditions C3 on each of the two-dimensional faces of the simplex. Choosing the coordinates of the nodes for specifying conditions C3 in the form

$$z_{i_1} = z_{i_2} = z_{i_3} = 1/3$$

we have the HIP satisfying this conditions C3 with one in r.h.s.

$$\check{\varphi}_{s''q}^{\kappa p'}(z) = 27 z_{i_1} z_{i_2} z_{i_3}.$$

2. Generalization of the fifth-order HIP basis $[p\kappa_{max}\kappa'_1 \cdots \kappa'_{d-1}] = [130 \cdots 0]$, $p = 1$, $\kappa_{max} = 3$, $p' = 5$. In addition to conditions C1, it contains three conditions C3 on each of the two-dimensional faces of the simplex $z_{i_1} = z_{i_2} = 1/4$, $z_{i_3} = 1/2$, four C3 on each of the three-dimensional faces of the simplex $z_{i_1} = z_{i_2} = z_{i_3} = 1/5$, $z_{i_4} = 2/5$, and one C3 on each of the four-dimensional faces of the simplex $z_{i_1} = z_{i_2} = z_{i_3} = z_{i_4} = z_{i_5} = 1/5$, the HIP satisfying this conditions C3 with one in right hand side have the form

$$
\begin{aligned}
\check{\varphi}_{s''q}^{\kappa p'}(z) &= -\frac{256}{25} z_{i_1} z_{i_2} z_{i_3} \\
&\quad \times \left(5(z_{i_1}^2 + z_{i_2}^2) - 45 z_{i_3}^2 + 40 z_{i_1} z_{i_2} + (z_{i_1} + z_{i_2})(1 - 60 z_{i_3}) + 51 z_{i_3} - 6 \right), \\
\check{\varphi}_{s''q}^{\kappa p'}(z) &= -\frac{625}{2} z_{i_1} z_{i_2} z_{i_3} z_{i_4} (5 z_{i_4} - 1), \\
\check{\varphi}_{s''q}^{\kappa p'}(z) &= 3125 z_{i_1} z_{i_2} z_{i_3} z_{i_4} z_{i_5},
\end{aligned}
\tag{3.43}
$$

respectively.

3.3.5   Hermite Interpolation Polynomials on Parallelepiped

The HIPs of d variables in d-dimensional hyperparallelepipeds are calculated in analytical form as an product of the d one dimensional HIPs $\varphi_{i_s q}^{\kappa_s p'}(z_s)$ depending on each of the d variables

$$\varphi_{i_1 \cdots i_d q}^{\kappa_1 \cdots \kappa_d p'}(z_1, \ldots, z_d) = \prod_{s=1}^{d} \varphi_{i_s q}^{\kappa_s p'}(z_s).
\tag{3.44}$$

It is easily proved that the HIPs on a parallelepiped are satisfy following conditions

$$\frac{\partial^{\kappa_1' + \cdots + \kappa_d'} \varphi_{i_1 \cdots i_d q}^{\kappa_1 \cdots \kappa_d p'}}{\partial z_1^{\kappa_1'} \cdots \partial z_d^{\kappa_d'}} (z_1', \ldots, z_d') = \delta_{z_1 z_1'} \cdots \delta_{z_d z_d'} \delta_{\kappa_1 \kappa_1'} \cdots \delta_{\kappa_d \kappa_d'}, \tag{3.45}$$

$$\kappa_s = 0, \ldots, \kappa_{\max} - 1.$$

Then a suitable way to construct these HIPs is a product (3.44) of the one dimensional HIPs calculated by recurrence relations (3.15)–(3.24), i.e. without solving a nonhomogeneous algebraic problem (3.42) as proposed in [35]. So, the HIPs on prismatic polytopes may be constructed by a similar way. For example, the basis of the HIPs on a prism, which will preserve the continuity of the first derivative at the boundaries, will consist of polynomials of the eighth order: the products of one-dimensional HIPs of the third order and the HIPs of the fifth order on a triangle.

3.3.6 Piecewise Polynomial Functions

The piecewise polynomial functions $N_l^{p'}(z) \equiv N_{tq}^{\kappa p'}(z)$

$$N_{tq}^{\kappa p'}(z_{t'}) = \delta_{tt'} \delta_{\kappa 0}, \quad \kappa = \kappa_1, \ldots, \kappa_d, \quad |\mu| = \mu_1 + \cdots + \mu_d,$$

$$\left. \frac{\partial^{|\mu|} N_{tq}^{\kappa p'}(z)}{\partial \tau^{|\mu|}} \right|_{z=z_{t'}} \equiv \left. \frac{\partial^{|\mu|} N_{tq}^{\kappa p'}(z)}{\partial \tau_1^{\mu_1} \cdots \partial \tau_d^{\mu_d}} \right|_{z=z_{t'}} = \delta_{tt'} \delta_{\kappa \mu}, \tag{3.46}$$

where τ are some directions determined by a finite element mesh $\Delta_q \in \Omega_h(z) = \bigcup_{q=1}^Q \Delta_q$: are constructed by joining the polynomials $\varphi_{lq}^{p'}(z) = \varphi_{tq}^{\kappa p'}(z)$:

$$N_{tq}^{\kappa p'}(z) = \left\{ LC_t^{|\kappa|} \varphi_{tq}^{\kappa p'}(z'), z_t \in \Delta_q; 0, z_t \notin \Delta_q \right\}, \quad |\kappa| = \kappa_1 + \cdots + \kappa_d, \tag{3.47}$$

where the $LC_t^{|\kappa|}$ means the linear combination of all HIPs obtained in the local coordinates, with the derivatives of order $|\kappa|$ equal to one in node z_t.

An example of piecewise polynomial functions constructed by joining the LIP [510] and HIP [131] of the fifth order shown in the Fig. 3.6. It can be seen from the figure that in the case of piecewise polynomial functions from the HIP, the isolines are smooth curves, which shows the continuity of the first-order derivatives.

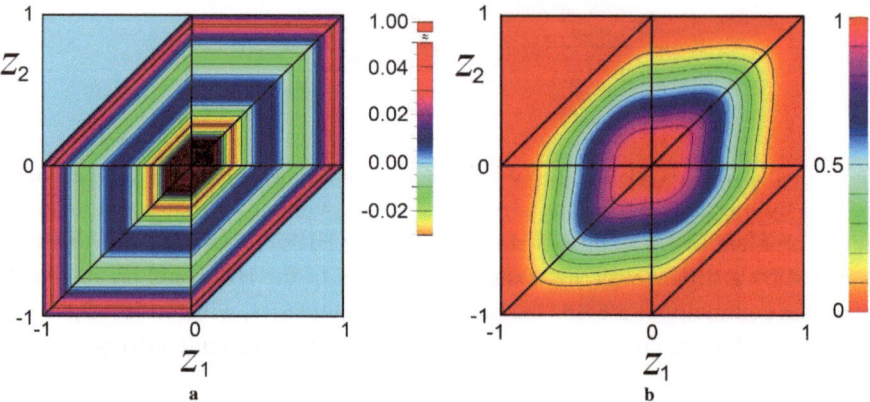

Fig. 3.6 Isolines of piecewise polynomial functions equal to one at the origin, obtained by stitching on neighboring finite elements—right-angled triangles with legs equal to 1: **a** fifth-order LIP and **b** fifth-order HIP (Argyris triangle [28])

3.4 Examples

3.4.1 The Helmholtz Problem on Triangle

As an example, let us consider the solution of the discrete-spectrum problem (3.1)–(3.5) at $d = 2$, $g_0(z) = g_{ij}(z) = \delta_{ij}$, and $V(z) = 0$ in the domain $\Omega_h(z) = \cup_{q=1}^{Q} \Delta_q$ in the form of an equilateral triangle with the side $4\pi/3$ with the boundary conditions of the second kind (3.4), partitioned into $Q = n^2$ equilateral triangles Δ_q with the side $h = 4\pi/3n$ [40].

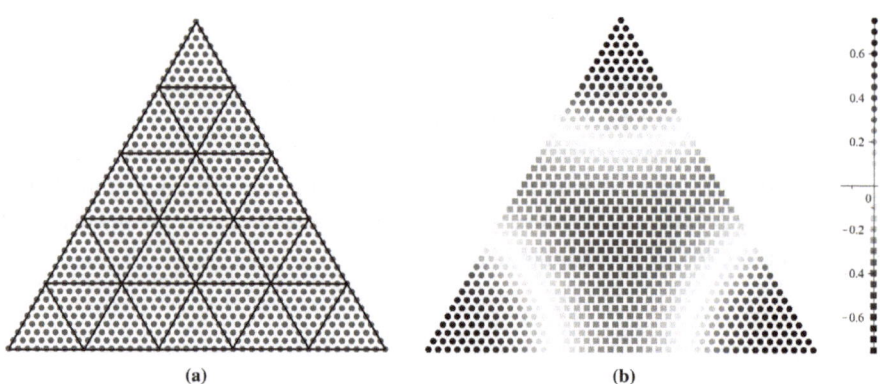

Fig. 3.7 a The mesh on the domain $\Omega_h(z) = \bigcup_{q=1}^{Q} \Delta_q$ of the triangle membrane, composed of triangle elements Δ_q. **b** the profiles of the fourth eigenfunction $\Phi_4^h(z)$ with $E_4^h = 3 + 1.90 \cdot 10^{-17}$, obtained using the LIP of the order $p' = p = 8$

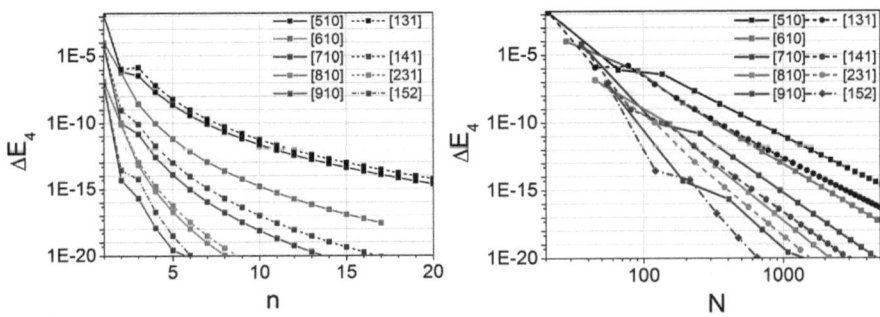

Fig. 3.8 The error ΔE_4^h of the eigenvalue E_4^h versus the number of elements n and the length of the vector N

The eigenvalues of this problem having the degenerate spectrum [69, 70] are the integers $E_m = m_1^2 + m_2^2 + m_1 m_2 = 0, 1, 1, 3, 4, 4, 7, 7, \ldots,$ $m_1, m_2 = 0, 1, 2, \ldots$. Figure 3.7 presents the finite element mesh with the LIP of the eighth-order and the profile of the fourth eigenfunction $\Phi_4^h(z)$. Figure 3.8 shows the errors $\Delta E_m = E_m^h - E_m$ of the eigenvalue $E_4^h(z)$ depending on the number n (the number of elements being n^2) and on the length N of the vector Φ_m^h of the algebraic eigenvalue problem for the FEM from the fifth to the ninth-order of accuracy: using LIP with the labels $[p\kappa_{\max}\kappa'] = [510], \ldots, [910]$, and using HIP with the labels $[p\kappa_{\max}\kappa'] = [131], [141], [231], [152]$ conserving the continuity of the first and the second derivative of the approximate solution, respectively. All numerical calculations were performed using Maple 18 system and quadruple-precision arithmetic on Intel FORTRAN Compiler.

As seen from Fig. 3.8, the errors of the eigenvalue $\Delta E_4^h(z)$ of the FEMs of the same order are nearly similar and correspond to the theoretical estimates (3.13), but in the FEMs conserving the continuity of the first and the second derivatives of the approximate solution the matrices of smaller dimension are used that correspond to the length of the vector N smaller by 1.5–2 times than in the methods with LIP that conserve only the continuity of the functions themselves at the boundaries of the finite elements.

3.4.2 The Helmholtz Problem on Hypercube

As examples of applying the above algorithms, we present the results of solving the Helmholtz problem (3.1)–(3.5) at $g_0(z) = g_{ij}(z) = \delta_{ij}$, and $V(z) = 0$ with Neumann boundary conditions for a multidimensional hypercube with the edge length π [71]. This problem has the degenerated spectrum with integer eigenvalues

$$E_m = \sum_{i=1}^{d} m_i^2, \tag{3.48}$$

and the well known eigenfunctions

$$\Phi_m(z) = \prod_{i=1}^{d} \Phi_{m_i}(z_i), \quad \Phi_{m_i}(z_i) = \frac{\sqrt{2 - \delta_{0m_i}}}{\sqrt{\pi}} \cos(m_i z_i). \tag{3.49}$$

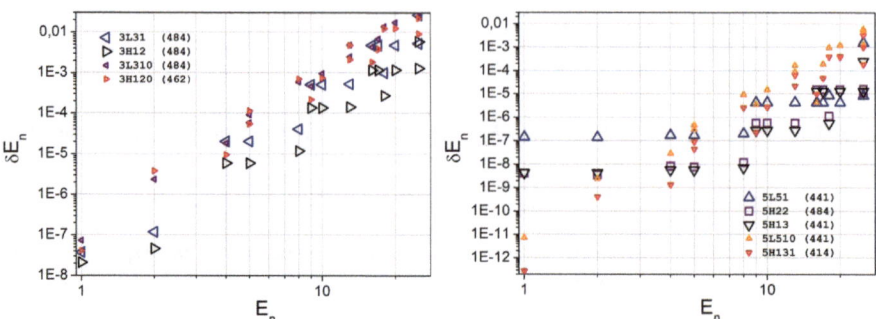

Fig. 3.9 The discrepancy $\delta E_m = E_m^h - E_m$ of calculated eigenvalue E_m^h of the Helmholtz problem for a square with the edge length π. Calculations were performed using FEM with third-order and fifth-order (3Ls and 5Ls) simplex Lagrange elements, and parallelepiped Lagrange (3Lp and 5Lp) and Hermite (3Hp and 5Hp) elements. The dimension of each eigenvectors of the algebraic problem is indicated in parentheses

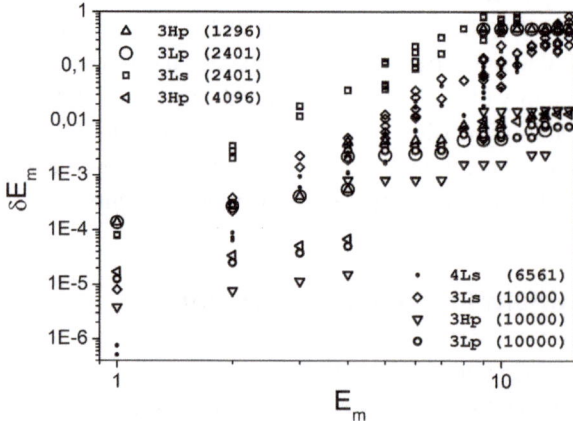

Fig. 3.10 The discrepancy $\delta E_m = E_m^h - E_m$ of calculated eigenvalue E_m^h of the Helmholtz problem for a four-dimensional cube with the edge length π. Calculations were performed using FEM with third-order (3Ls) and fourth-order (4Ls) simplex Lagrange elements, and third-order parallelepiped Lagrange (3Lp) and Hermite (3Hp) elements. The dimension of eigenvectors of the algebraic problem is given in parentheses

Figure 3.9 shows the discrepancy $\delta E_m = E_m^h - E_m$ between the numerical eigenvalues E_m^h of the 2D problem and their exact values $E_m = 0_1, 1_2, 2_1, 4_2, 5_2, \ldots$, where subscript means the multiplicity of degeneracy. The results were calculated using FEM with the square LIPs and HIPs of third and fifth-order that are obtained by multiplying two 1D LIPs or two 1D HIPs, respectively, in accordance (3.44). As one would expect with a fairly small step, the higher-order methods demonstrate better performance in comparison with the low-order ones, and the methods with HIPs work better than those using LIPs of the same order.

Figure 3.10 shows the discrepancy $\delta E_m = E_m^h - E_m$ between the numerical eigenvalues E_m^h of the 4D problem and their exact values $E_m = 0_1, 1_4, 2_6, 3_4, 4_5, 5_{12}, \cdots$. The results were calculated using FEM with hypercube LIPs and HIPs of the third-order that are obtained by product of four 1D LIPs or four 1D HIPs, respectively, in accordance (3.44). They are compared with simplex LIPs of the third- and the fourth-order. There is a stepwise structure of the discrepancy δE_m calculated with 4D hypercubic LIPs and HIPs, with the steps appearing at the values $E_m = 1, 4, 9, \ldots$, which is similar to the stepwise structure shown in Fig. 3.9. The structure is also due to the prevalence of approximation errors of eigenfunctions caused by the pure partial derivatives. For the simplex LIPs the oscillating structure of the discrepancy δE_m is due to different contributions the approximation errors caused by the different mixed partial derivatives.

3.4.3 Quadrupole-Octupole-Vibrational Collective Model

Below we solve the BVP (3.1)–(3.5) in the 2D domain ($d = 2$) that describe the quadrupole-octupole-vibrational collective model of ^{156}Dy nucleus [7, 72] with the coefficients given in tabular form $g_0(z)$ and $g_{ij}(z)$ determined by the expressions $i, j = 1, 2$:

$$g_0(z_1, z_2) = \frac{2}{\hbar^2} \sqrt{\det B(z_1, z_2)},$$
$$g_{ij}(z_1, z_2) = \sqrt{\det B(z_1, z_2)} [B^{-1}(z_1, z_2)]_{ij}. \tag{3.50}$$

The mass tensor $B_{ij}(z_1, z_2)$ and the potential energy function $V(z_1, z_2)$ has been calculated [72] in the terms of the average nuclear deformations $z = (z_1, z_2) = (q_{20}, q_{32})$ determined in [73], and shown in Figs. 3.11 and 3.12a.

Table 3.4 shows a low part of the spectrum of $v = 1, \ldots, 10$ states of ^{156}Dy counted from minimum of potential energy ($V_{\min}(z_1, z_2) = 0.685$ MeV). Second column shows eigenenergies E_v^{FDM} calculated by the FDM code of the second-order [72]. The remaining columns show the eigenvalues $E_v^{\text{FEM}}(p)$ of the BVP (3.1)–(3.5) in $\Omega(z_1, z_2)$ with coefficients $g_{ij}(z_1, z_2)$ determined by formulas (3.5) and the potential energy functions $V(z_1, z_2)$ calculated by the FEM code with the PI-type Gaussian quadrature rules till the eight-order (see Appendix B). Calculations has been carried out with the second type boundary conditions (3.4) and orthonormalization condition (3.5) with triangular Lagrange elements of

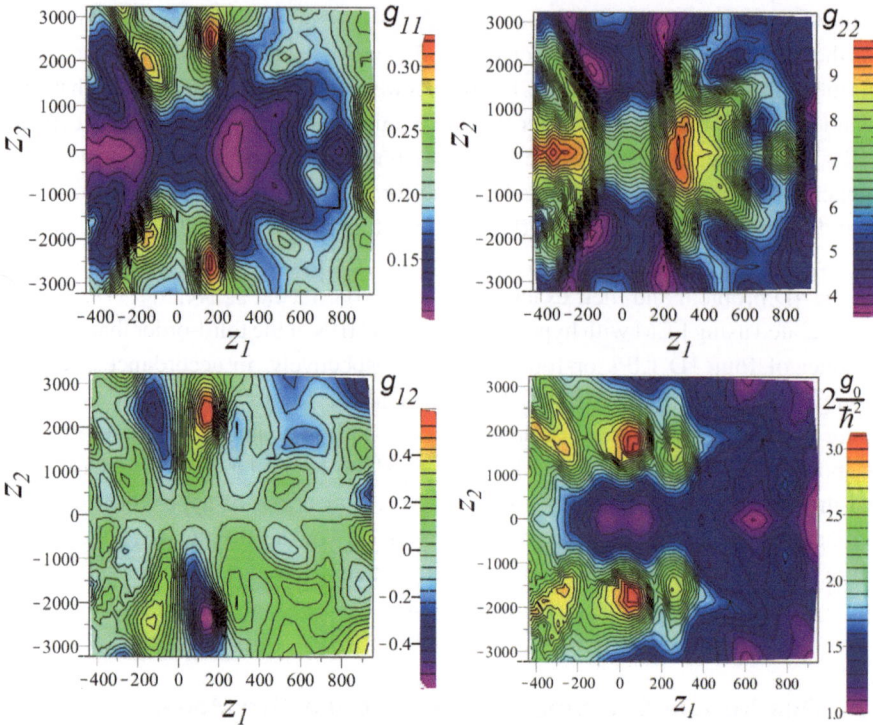

Fig. 3.11 The coefficients $g_{ij}(z)$ from (3.50) and square root of the determinant $g_0(z)$ constructed out of collective inertia parameters in units $10^{-5}\hbar^2/(\text{MeV fm}^5)$

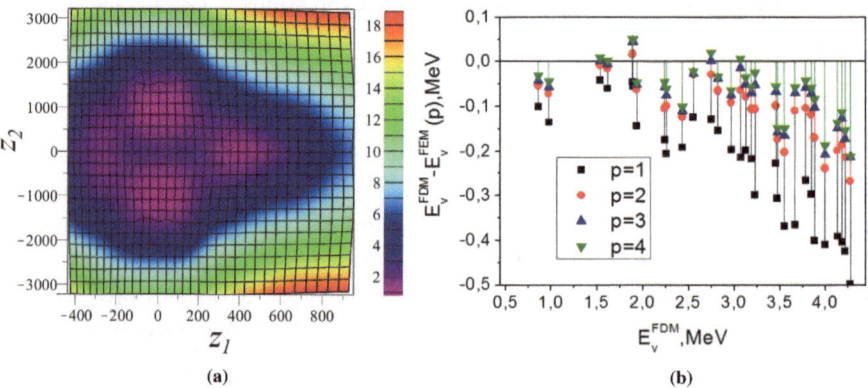

(a)   (b)

Fig. 3.12 **a** The potential energy $V(z_1, z_2)$ of ^{156}Dy nucleus given in variables $(z_1, z_2) = (q_{20}, q_{32})$. The nodal points of finite element grid are intersection points of horizontal and vertical lines. **b** The differences $E_v^{\text{FDM}} - E_v^{\text{FEM}}(p)$ between eigenvalues of E_v^{FDM} of ^{156}Dy nucleus calculated by the FDM [72] and $E_v^{\text{FEM}}(p)$ calculated by FEM with triangular Lagrange elements of the order $p = 1, 2, 3, 4$ for 30 lowest states of the BVP (3.1)–(3.5) in variables $(z_1, z_2) = (q_{20}, q_{32})$

Table 3.4 The low part of the spectrum of 10 lowest states of ^{156}Dy nucleus counted from minimum of potential energy ($V_{\min}(z_1, z_2) = 0.685$ MeV). E_v^{FDM} calculated by FDM of the second-order [72] and $E_v^{\text{FEM}}(p)$ calculated by FEM with triangular Lagrange elements of the order $p = 1, 2, 3, 4$ in [7]

v	E_v^{FDM}	$E_v^{\text{FEM}}(1)$	$E_v^{\text{FEM}}(2)$	$E_v^{\text{FEM}}(3)$	$E_v^{\text{FEM}}(4)$
1	0.85988	0.96000	0.91329	0.90234	0.89065
2	0.97588	1.11144	1.04808	1.03297	1.02068
3	1.53669	1.57813	1.54403	1.53371	1.52776
4	1.61774	1.67776	1.63332	1.62287	1.61571
5	1.88907	1.93560	1.87335	1.84504	1.83794
6	1.89469	1.94932	1.87706	1.84925	1.84631
7	1.93369	2.07731	1.99714	1.98486	1.98032
8	2.23907	2.41405	2.34335	2.29594	2.28444
9	2.25778	2.46383	2.35681	2.33287	2.31778
10	2.43288	2.62454	2.55679	2.54278	2.53388

the order $p = 1, 2, 3, 4$ in the finite element grid $\Omega(z_1, z_2)$. Discrepancy $E_v^{\text{FDM}} - E_v^{\text{FEM}}(p)$ between the results of FDM and FEM calculations in dependence of the order $p = 1, 2, 3, 4$ of the FEM approximation is shown in Fig. 3.12b. One can see that in increasing the order of the FEM approximation the discrepancy is decreased till 1%. Figure 3.13 displays the corresponding isolines of FEM eigenfunctions $\Phi_v(z_1, z_2)$ calculated in the finite element grid $\Omega(z_1, z_2)$ [7]. The FEM eigenfunctions of the ground and first excited states are in good agreement with the eigenfunctions calculated in domain $\Omega(z_1, z_2)$ by the FDM [72]. The third FEM eigenfunction has one node line in direction of z_1 in contrast with the third FDM eigenfunction that has no nodes. Meanwhile, the fourth FEM function has two node lines in direction of z_1 and qualitative coincides with the fourth FDM eigenfunction. We can suppose that the revivable distinctions are consequence of approximation of table values of $V(z_1, z_2)$ on the FEM grid $\Omega(z_1, z_2)$ instead of approximation of derivatives of table values of $g_{ij}(z_1, z_2)$ on the FDM grid $\Omega(z_1, z_2)$ in [72].

Note that, the latter two-dimensional BVP has been solved using FDM [72] and FEM [7] that was only a part of the BVP in the 6D domain, where the potential energy and components of the metric tensor are given by 2×10^6 table values [6]. However, the FDM approach [72] did not obtain further generalization on the above multidimensional domain, while the elaborated FEM and PI-type Gaussian quadrature rules [66] have no restriction for further solving such BVP in the 6D domain. This obstacle has been one line of motivations in development and implementation of the above approach.

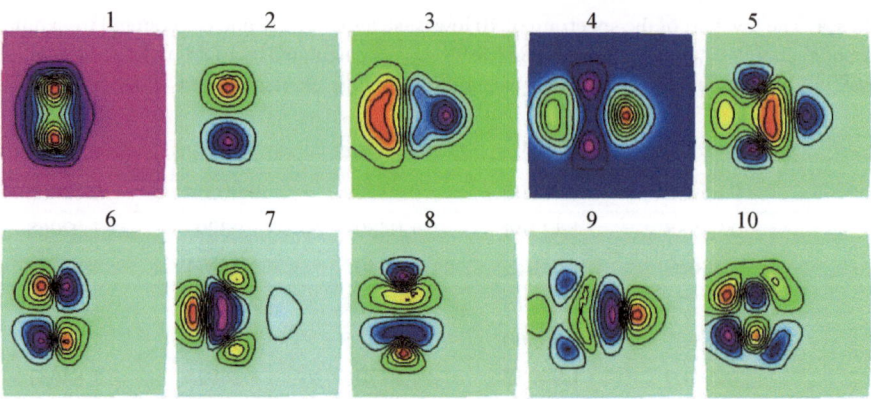

Fig. 3.13 The first ten FEM eigenfunctions of ^{156}Dy nucleus in the plane $(z_1, z_2) = (q_{20}, q_{32})$

References

1. S. Cwiok, J. Dudek, W. Nazarewicz, J. Skalski, T. Werner, Single-particle energies, wave functions, quadrupole moments and g-factors in an axially deformed Woods-Saxon potential with applications to the two-centre-type nuclear problems. Comput. Phys. Commun. **46**, 379–399 (1987)
2. O. Chuluunbaatar, A. Gusev, V. Gerdt, M. Kaschiev, V. Rostovtsev, V. Samoylov, T. Tupikova, S. Vinitsky, A symbolic-numerical algorithm for solving the eigenvalue problem for a hydrogen atom in the magnetic field: cylindrical coordinates. Lect. Notes Comput. Sci. **4770**, 118–133 (2007)
3. A.A. Gusev, O. Chuluunbaatar, V.P. Gerdt, V.A. Rostovtsev, S.I. Vinitsky, V.L. Derbov, V.V. Serov, Symbolic-numeric algorithms for computer analysis of spheroidal quantum dot models. Lect. Notes Comput. Sci. **6244**, 106–122 (2010)
4. A.A. Gusev, S.I. Vinitsky, O. Chuluunbaatar, V.P. Gerdt, V.A. Rostovtsev, Symbolic-numerical algorithms to solve the quantum tunneling problem for a coupled pair of ions. Lect. Notes Comput. Sci. **6885**, 175–191 (2011)
5. A. Gusev, S. Vinitsky, O. Chuluunbaatar, V.A. Rostovtsev, L.L. Hai, V. Derbov, P. Krassovitskiy, Symbolic-numerical algorithm for generating cluster eigenfunctions: quantum tunneling of clusters through repulsive barriers. Lect. Notes Comput. Sci. **8136**, 427–442 (2013)
6. A. Dobrowolski, K. Mazurek, A. Góźdź, Consistent quadrupole-octupole collective model. Phys. Rev. C **94**, 0543220–1–20 (2016)
7. A.A. Gusev, S.I. Vinitsky, O. Chuluunbaatar, A. Góźdź, A. Dobrowolski, K. Mazurek, P.M. Krassovitskiy, Finite element method for solving the collective nuclear model with tetrahedral symmetry. Acta Physica Polonica B Proc. Suppl. **12**, 589–594 (2019)
8. G. Strang, G.J. Fix, *An Analysis of the Finite Element Method* (Prentice-Hall, Englewood Cliffs, 1973)
9. E.B. Becker, G.F. Carey, J. Tinsley Oden, *Finite Elements. An Introduction*, vol. I (Prentice-Hall, Inc., Englewood Cliffs, New Jersey, 1981)
10. K.J. Bathe, *Finite Element Procedures in Engineering Analysis* (Englewood Cliffs, Prentice Hall, New York, 1982)

11. O. Chuluunbaatar, A.A. Gusev, V.L. Derbov, M.S. Kaschiev, L.A. Melnikov, V.V. Serov, S.I. Vinitsky, Calculation of a hydrogen atom photoionization in a strong magnetic field by using the angular oblate spheroidal functions. J. Phys. A **40**, 11485–11524 (2007)

12. A.A. Gusev, V.P. Gerdt, O. Chuluunbaatar, G. Chuluunbaatar, S.I. Vinitsky, V.L. Derbov, A. Góźdź, Symbolic-numerical algorithms for solving the parametric self-adjoint 2D elliptic boundary-value problem using high-accuracy finite element method. Lect. Notes Comput. Sci. **10490**, 151–166 (2017)

13. O. Chuluunbaatar, A.A. Gusev, A.G. Abrashkevich, A. Amaya-Tapia, M.S. Kaschiev, S.Y. Larsen, S.I. Vinitsky, KANTBP: a program for computing energy levels, reaction matrix and radial wave functions in the coupled-channel hyperspherical adiabatic approach. Comput. Phys. Commun. **177**, 649–675 (2007)

14. O. Chuluunbaatar, A.A. Gusev, S.I. Vinitsky, A.G. Abrashkevich, KANTBP 2.0: new version of a program for computing energy levels, reaction matrix and radial wave functions in the coupled-channel hyperspherical adiabatic approach. Comput. Phys. Commun. **179**, 685–693 (2008)

15. A.A. Gusev, O. Chuluunbaatar, S.I. Vinitsky, A.G. Abrashkevich, KANTBP 3.0: new version of a program for computing energy levels, reflection and transmission matrices, and corresponding wave functions in the coupled-channel adiabatic approach. Comput. Phys. Commun. **185**, 3341–3343 (2014)

16. O. Chuluunbaatar, A.A. Gusev, S.I. Vinitsky, A.G. Abrashkevich, P.W. Wen, C.J. Lin, KANTBP 3.1: a program for computing energy levels, reflection and transmission matrices, and corresponding wave functions in the coupled-channel and adiabatic approaches. Comput. Phys. Commun. **278**, 108397–1–14 (2022)

17. O. Chuluunbaatar, A.A. Gusev, S.I. Vinitsky, A.G. Abrashkevich, ODPEVP: a program for computing eigenvalues and eigenfunctions and their first derivatives with respect to the parameter of the parametric self-adjoined Sturm-Liouville problem. Comput. Phys. Commun. **180**, 1358–1375 (2009)

18. A.A. Gusev, O. Chuluunbaatar, S.I. Vinitsky, A.G. Abrashkevich, POTHEA: a program for computing eigenvalues and eigenfunctions and their first derivatives with respect to the parameter of the parametric self-adjoined 2D elliptic partial differential equation. Comput. Phys. Commun. **185**, 2636–2654 (2014)

19. O. Chuluunbaatar, A.A. Gusev, V.P. Gerdt, V.A. Rostovtsev, S.I. Vinitsky, A.G. Abrashkevich, M.S. Kaschiev, V.V. Serov, POTHMF: a program for computing potential curves and matrix elements of the coupled adiabatic radial equations for a hydrogen-like atom in a homogeneous magnetic field. Comput. Phys. Commun. **178**, 301–330 (2008)

20. G. Chuluunbaatar, A.A. Gusev, O. Chuluunbaatar, S.I. Vinitsky, L.L. Hai, KANTBP 4M program for solving the scattering problem for a system of ordinary second-order differential equations. EPJ Web Conf. **226**, 02008–1–4 (2020)

21. G. Chuluunbaatar, A. Gusev, V. Derbov, S. Vinitsky, O. Chuluunbaatar, L.L. Hai, V. Gerdt, A Maple implementation of the finite element method for solving boundary-value problems for systems of second-order ordinary differential equations. Commun. Comput. Inf. Sci. **1414**, 152–166 (2021)

22. A.A. Gusev, L.Le Hai, O. Chuluunbaatar, S.I. Vinitsky, KANTBP 4M: program for solving boundary problems of the system of ordinary second order differential equations. Program Library JINRLIB. http://wwwinfo.jinr.ru/programs/jinrlib/kantbp4m/indexe.html

23. S.I. Vinitsky, A.A. Gusev, O. Chuluunbaatar, TIME6T: program complex for the numerical solution of the Cauchy problem for the time-dependent Schroedinger equation. Program Library JINRLIB. http://wwwinfo.jinr.ru/programs/jinrlib/time6t/indexe.html

24. G. Chuluunbaatar, O. Chuluunbaatar, A.A. Gusev, S.I. Vinitsky, INQSIM: a program for converting PI-type fully symmetric quadrature rules on 2-, …, 6-simplexes from compact to expanded forms. Program Library JINRLIB. http://wwwinfo.jinr.ru/programs/jinrlib/inqsim/indexe.html

25. I.S. Berezin, N.P. Zhidkov, *Computing Methods*, vol. I (Pergamon Press, Oxford, 1965)

26. A.A. Samarskii, A.V. Goolin, *Numerical Methods* (Nauka, Moscow, 1989). ((in Russian))

27. W.F. Ames, *Numerical Methods for Partial Differential Equations* (Academic, London, 1992)

28. J.H. Argyris, K.E. Buck, D.W. Scharpf, H.M. Hilber, G. Mareczek, Some new elements for the matrix displacement method, in *Proceedings of the Conference on Matrix Methods in Structural Mechanics (2nd)*. Wright-Patterson Air Force Base, Ohio, 15–17 October 1968

29. K. Bell, A refined triangular plate bending element. Int. J. Numer. Methods Eng. **1**, 101–122 (1969)

30. S.C. Brenner, L.R. Scott, *The Mathematical Theory of Finite Element Methods*. Texts in Applied Mathematics, 3rd edn., vol. 15 (Springer, New York, 2008)

31. P. Ciarlet, *The Finite Element Method for Elliptic Problems* (North-Holland Publishing Company, Amsterdam, 1978)

32. G. Dhatt, G. Touzot, E. Lefrançois, *Finite Element Method* (Wiley, 2012)

33. M. Gasca, T. Sauer, On the history of multivariate polynomial interpolation. J. Comp. Appl. Math. **122**, 23–35 (2000)

34. A.W. Habib, R.N. Goldman, T. Lyche, A recursive algorithm for Hermite interpolation over a triangular grid. J. Comput. Appl. Math. **73**, 95–118 (1996)

35. F. Lekien, J. Marsden, Tricubic interpolation in three dimensions. Int. J. Numer. Methods Eng. **63**, 455–471 (2005)

36. A. Logg, K.-A. Mardal, G.N. Wells (eds.), *Automated Solution of Differential Equations by the Finite Element Method (The FEniCS Book)* (Springer, Berlin, Heidelberg, 2012)

37. A.R. Mitchell, R. Wait, *The Finite Element Method in Partial Differential Equations* (Wiley, Chichester, 1977)

38. O.C. Zienkiewicz, Finite elements - The background story, in *The Mathematics of Finite Elements and Applications*. ed. by J.R. Whiteman (Academic, London, 1973), pp.1–35

39. A.A. Gusev, O. Chuluunbaatar, S.I. Vinitsky, V.L. Derbov, A. Góźdź, L.L. Hai, V.A. Rostovtsev, Symbolic-numerical solution of boundary-value problems with self-adjoint second-order differential equation using the finite element method with interpolation Hermite polynomials. Lect. Notes Comput. Sci. **8660**, 138–154 (2014)

40. A.A. Gusev, V.P. Gerdt, O. Chuluunbaatar, G. Chuluunbaatar, S.I. Vinitsky, V.L. Derbov, A. Góźdź, Symbolic-numerical algorithm for generating interpolation multivariate Hermite polynomials of high-accuracy finite element method. Lect. Notes Comput. Sci. **10490**, 134–150 (2017)

41. M. Festa, A. Sommariva, Computing almost minimal formulas on the square. J. Comput. Appl. Math. **236**, 4296–4302 (2012)

42. E.K. Ryu, S.P. Boyd, Extensions of Gauss quadrature via linear programming. Found. Comput. Math. **15**, 953–971 (2015)

43. S. Jayan, K.V. Nagaraja, Generalized Gaussian quadrature rules over regions with parabolic edges. Int. J. Comput. Math. **89**, 1631–1640 (2012)

44. P.C. Hammer, O.J. Marlowe, A.H. Stroud, Numerical integration over simplexes and cones. Math. Tabl. Natn. Rex Coun. Wash. **10**, 130–137 (1956)

45. P.C. Hammer, A.H. Stroud, Numerical integration over simplexes. Math. Tabl. Natn. Rex Coun. Wash. **10**, 137–139 (1956)

46. A. Grundmann, H.M. Möller, Invariant integration formulas for the n-simplex by combinatorial methods. SIAM J. Numer. Anal. **15**, 282–290 (1978)

47. P. Silvester, Symmetrie quadrature formulae for simplexes. Math. Comp. **24**, 95–100 (1970)

48. G.R. Cowper, Gaussian quadrature formulas for triangles. Int. J. Numer. Methods Eng. **7**, 405–408 (1973)
49. J.N. Lyness, D. Jespersen, Moderate degree symmetric quadrature rules for the triangle. J. Inst. Maths. Applies **15**, 19–32 (1975)
50. M.E. Laursen, M. Gellert, Some criteria for numerically integrated matrices and quadrature formulas for triangles. Int. J. Numer. Methods Eng. **12**, 67–76 (1978)
51. D.A. Dunavant, High degree efficient symmetrical Gaussian quadrature rules for the triangle. Int. J. Numer. Methods Eng. **21**, 1129–1148 (1985)
52. L. Zhang, T. Cui, H. Liu, A set of symmetric quadrature rules on triangles and tetrahedra. J. Comput. Math. **27**, 89–96 (2009)
53. M.A. Taylor, B.A. Wingate, L.P. Bos, Several new quadrature formulas for polynomial integration in the triangle (2007), pp. 1–14. arXiv:math/0501496
54. S. Wandzura, H. Xiao, Symmetric quadrature rules on a triangle. Comput. Math. Appl. **45**, 1829–1840 (2003)
55. H. Xiao, Z. Gimbutas, A numerical algorithm for the construction of efficient quadrature rules in two and higher dimensions. Comput. Math. Appl. **59**, 663–676 (2010)
56. D.M. Williams, L. Shunn, A. Jameson, Symmetric quadrature rules for simplexes based on sphere close packed lattice arrangements. J. Comput. Appl. Math. **266**, 18–38 (2014)
57. F.D. Witherden, P.E. Vincent, On the identification of symmetric quadrature rules for finite element methods. Comput. Math. Appl. **69**, 1232–1241 (2015)
58. B.A. Freno, W.A. Johnson, B.F. Zinser, S. Campione, Symmetric triangle quadrature rules for arbitrary functions. Comput. Math. Appl. **79**, 2885–2896 (2020)
59. A.A. Gusev, V.P. Gerdt, O. Chuluunbaatar, G. Chuluunbaatar, S.I. Vinitsky, V.L. Derbov, A. Góźdź, P.M. Krassovitskiy, Symbolic-numerical algorithms for solving elliptic boundary-value problems using multivariate simplex Lagrange elements. Lect. Notes Comput. Sci. **11077**, 197–213 (2018)
60. S. Geevers, W.A. Mulder, J.J.W. Van Der Vegt, Efficient quadrature rules for computing the stiffness matrices of mass-lumped tetrahedral elements for linear wave problems. Siam J. Sci. Comput. **41**, A1041–A1065 (2019)
61. J. Jaśkowiec, N. Sukumar, High-order cubature rules for tetrahedra. Int. J. Numer. Methods Eng. **121**, 2418–2436 (2020)
62. J. Jaśkowiec, N. Sukumar, High-order symmetric cubature rules for tetrahedra and pyramids. Int. J. Numer. Methods Eng. **122**, 148–171 (2021)
63. E. Sainz de la Maza, Fórmulas de cuadratura invariantes de grado 8 para el simplex 4-dimensional. Revista Internacional de Métodos Numéricos para Cálculo y Diseño en Ingeniería **15**, 375–379 (1999)
64. D.M. Williams, C.V. Frontin, E.A. Miller, D.L. Darmofal, A family of symmetric, optimized quadrature rules for pentatope. Comput. Math. Appl. **80**, 1405–1420 (2020)
65. C.V. Frontin, G.S. Walters, F.D. Witherden, W. Lee, D.M. Williams, D.L. Darmofal, Foundations of space-time finite element methods: polytopes, interpolation, and integration. Appl. Numer. Math. **166**, 92–113 (2021)
66. G. Chuluunbaatar, O. Chuluunbaatar, A.A. Gusev, S.I. Vinitsky, PI-type fully symmetric quadrature rules on the 3-, ..., 6-simplexes. Comput. Math. Appl. **124**, 89–97 (2022)
67. O.A. Ladyzhenskaya, *The Boundary Value Problems of Mathematical Physics*. Applied Mathematical Sciences 49 (Springer, New York, 1985)
68. A. Ženíšek, Hermite interpolation on simplexes in the finite element method, in *Proceedings of Equadiff III, 3rd Czechoslovak Conference on Differential Equations and Their Applications*. Brno, Czechoslovakia, August 28–September 1, 1972. V. 1 (Brno, 1973), pp. 271–277

69. B.J. McCartin, *Laplacian Eigenstructure of the Equilateral Triangle* (Hikari Ltd, Ruse, Bulgary, 2011)
70. F. Pockels, *Über die partielle differential-gleichung* $\Delta u + k^2 u = 0$ *und deren auftreten in der mathematischen physik* (B.G. Teubner, Leipzig, 1891)
71. G. Chuluunbaatar, A.A. Gusev, O. Chuluunbaatar, V.P. Gerdt, S.I. Vinitsky, V.L. Derbov, A. Góźdź, P.M. Krassovitskiy, L.L. Hai, Construction of multivariate interpolation Hermite polynomials for finite element method. EPJ Web Conf. **226**, 02007–1–4 (2020)
72. A. Dobrowolski, K. Mazurek, J. Dudek, Tetrahedral symmetry in nuclei: New predictions based on the collective model. Int. J. Mod. Phys. E **20**, 500–506 (2011)
73. A. Dobrowolski, H. Goutte, J.-F. Berger, Microscopic determinations of fission barriers (mean-field and beyond). Int. J. Mod. Phys. E **16**, 431–442 (2007)

Continuous Analogue of Newton's Method for Solving the Generalized Eigenvalue Problem

We apply the continuous analogue of Newton's method for solving the generalized eigenvalue problem [1, 2]

$$(\mathbf{A} - \lambda \mathbf{B}) \mathbf{u} = 0, \tag{A.1}$$

$$(\mathbf{u}, \mathbf{Bu}) = 1, \tag{A.2}$$

with respect to the unknown pair $\{\lambda, \mathbf{u}\}$. Here $\mathbf{A}$ and $\mathbf{B}$ are real square matrices, $\mathbf{B}$ is positive dependent matrix. The evolution equations corresponding to the problems (A.1), (A.2) have the form

$$(\mathbf{A} - \lambda \mathbf{B}) \frac{d\mathbf{u}}{dt} - \frac{d\lambda}{dt} \mathbf{Bu} = -(\mathbf{A} - \lambda \mathbf{B}) \mathbf{u}, \tag{A.3}$$

$$2 \left(\frac{d\mathbf{u}}{dt}, \mathbf{Bu} \right) = 1 - (\mathbf{u}, \mathbf{Bu}). \tag{A.4}$$

Using the discrete approximation of the first derivatives

$$\frac{d\mathbf{u}}{dt}\bigg|_{t_n} \approx \frac{\mathbf{u}_{n+1} - \mathbf{u}_n}{\tau_n} = \mathbf{v}_n, \quad \frac{d\lambda}{dt}\bigg|_{t_n} \approx \frac{\lambda_{n+1} - \lambda_n}{\tau_n} = \mu_n, \tag{A.5}$$

we arrive it to the iterative scheme

$$(\mathbf{A} - \lambda_n \mathbf{B}) \mathbf{v}_n - \mu_n \mathbf{Bu}_n = -\mathbf{r}_n, \tag{A.6}$$

$$2 (\mathbf{v}_n, \mathbf{Bu}_n) = 1 - (\mathbf{u}_n, \mathbf{Bu}_n), \tag{A.7}$$

where $\mathbf{r}_n$ is an residual of Eq. (A.1)

$$\mathbf{r}_n = (\mathbf{A} - \lambda_n \mathbf{B}) \mathbf{u}_n, \tag{A.8}$$

U. Vandandoo et al., *High-Order Finite Difference and Finite Element Methods for Solving Some Partial Differential Equations*, Synthesis Lectures on Engineering, Science, and Technology, https://doi.org/10.1007/978-3-031-44784-6

If $\mathbf{v}_n$ is represented in the form

$$\mathbf{v}_n = -\mathbf{u}_n + \mu_n \mathbf{\Theta}_n, \tag{A.9}$$

where $\mathbf{\Theta}_n$ is the solution to the problem

$$(\mathbf{A} - \lambda_n \mathbf{B})\, \mathbf{\Theta}_n = \mathbf{B} \mathbf{u}_n, \tag{A.10}$$

from (A.7) we have the following expression for μ_n

$$\mu_n = \frac{1 + (\mathbf{u}_n, \mathbf{B} \mathbf{u}_n)}{2\,(\mathbf{\Theta}_n, \mathbf{B} \mathbf{u}_n)}. \tag{A.11}$$

Thus, we obtain the following expression for new approximations:

$$\mathbf{u}_{n+1} = \mathbf{u}_n + \tau_n \mathbf{v}_n = (1 - \tau_n)\mathbf{u}_n + \tau_n \mu_n \mathbf{\Theta}_n, \tag{A.12}$$
$$\lambda_{n+1} = \lambda_n + \tau_n \mu_n.$$

The residual (A.8) at the $(n+1)$th iteration has form:

$$\mathbf{r}_{n+1} = (1 - \tau_n)\mathbf{r}_n - \tau_n^2 \mu_n \mathbf{B} \mathbf{v}_n \tag{A.13}$$

and

$$\|\mathbf{r}_{n+1}\|^2 = (1 - \tau_n)^2 \|\mathbf{r}_n\|^2 - 2(1 - \tau_n)\tau_n^2 \mu_n (\mathbf{r}_n, \mathbf{B} \mathbf{v}_n) + \tau_n^4 \mu_n^2 \|\mathbf{B} \mathbf{v}_n\|^2. \tag{A.14}$$

We denote $f(\tau_n) = \|\mathbf{r}_{n+1}\|^2$, consider its first derivative

$$f'(\tau_n) = 2(\tau_n - 1)\|\mathbf{r}_n\|^2 - 2(2\tau_n - 3\tau_n^2)\mu_n (\mathbf{r}_n, \mathbf{B} \mathbf{v}_n) + 4\tau_n^3 \mu_n^2 \|\mathbf{B} \mathbf{v}_n\|^2. \tag{A.15}$$

One can see that $f'(0) = -2\|\mathbf{r}_n\|^2 < 0$ and $f'(2) = 2f(2) > 0$. It means that the function $f'(\tau_n)$ has at least one root on the interval $(0, 2)$, i.e., the function $f(\tau_n)$ has at least a minimum. If it has three real roots, then the $f(\tau_n)$ has minimums at the smallest and largest roots and should choose the root that give the smallest value. Thus it is possible to find the optimum value iteration parameter

$$\tau_n = \tau_n^{\text{opt}}, \tag{A.16}$$

via calculating cubic equation $f'(\tau_n^{\text{opt}}) = 0$ (may be useful to use Cardano's formula).

In [3] given another optimal parameter τ_n

$$\tau_n^{\text{kal}} = \frac{\delta_n}{\delta_n + \delta_{n+1}(1)} \in (0, 1], \tag{A.17}$$

where $\delta_n = \|\mathbf{r}_{n+1}\|^2$, and $\delta_{n+1}(1) = \|\hat{\mathbf{r}}_{n+1}\|^2$ is the residual at the $(n+1)$th iteration for $\tau_n = 1$.

References

1. I.V. Puzynin, T.L. Boyadjiev, S.I. Vinitsky, E.V. Zemlyanaya, T.P. Puzynina, O. Chu-luunbaatar, Methods of computational physics for investigation of models of complex physical systems, Phys. Part. Nucl. **38**, 70–116 (2007)
2. T. Zhanlav, R. Mijiddorj, O. Chuluunbaatar, The continuous analogue of Newton's method for solving eigenvalues and eigenvectors of matrices, Bull. Bull. Tver State Univ. **14**, 27–37 (2008) (in Russian)
3. V.V. Ermakov, N.N. Kalitkin, Optimal step and regulation of Newton's method, Zh. Vych. Mat. Mat. Fiz. **21**, 491–497 (1981)

PI-Type Fully Symmetric Quadrature Rules on the Simplexes

B.1 Construction of Fully Symmetric Quadrature Rules

Let us construct the p-order quadrature rule

$$\int_{\Delta_d} V(x)dx = \frac{1}{d!} \sum_{j=1}^{N_{dp}} w_j V(x_{j1}, \ldots, x_{jd}), \tag{B.1}$$

$$x = (x_1, \ldots, x_d), \quad dx = dx_1 \cdots dx_d,$$

for integration over the standard unit d-simplex Δ_d with vertices $\hat{x}_j = (\hat{x}_{j1}, \ldots, \hat{x}_{jd})$, $\hat{x}_{jk} = \delta_{jk}$, $j = 0, \ldots, d$, $k = 1, \ldots, d$, which is exact for all polynomials of the variables $x_1, \ldots, x_d$ of degree not exceeding p. In Eq. (B.1), N_{dp} is the number of nodes, w_j are the weights, and $(x_{j1}, \ldots, x_{jd})$ are the nodes.

To building up PI-type fully symmetric quadrature rules, we use the barycentric coordinates (BC) $(y_1, \ldots, y_{d+1})$ of nodes:

$$\sum_{k=1}^{d+1} y_k = 1. \tag{B.2}$$

Using the invariance of the symmetric quadrature rules, Eq. (B.1) can be represented in the symmetric expanded form,

$$\int_{\Delta_d} V(x)dx = \frac{1}{d!} \sum_{j=1}^{N} w_j \sum_{k_1, \ldots, k_{d+1}} V(y_{jk_1}, \ldots, y_{jk_d} | y_{jk_{d+1}}), \tag{B.3}$$

© The Editor(s) (if applicable) and The Author(s), under exclusive license to Springer Nature Switzerland AG 2024
U. Vandandoo et al., *High-Order Finite Difference and Finite Element Methods for Solving Some Partial Differential Equations*, Synthesis Lectures on Engineering, Science, and Technology, https://doi.org/10.1007/978-3-031-44784-6

where the internal summation over $k_1, \ldots, k_{d+1}$ is carried out over the different permutations of the BC $(y_{j1}, \ldots, y_{jd+1})$, while $V(y_{jk_1}, \ldots, y_{jk_d}|y_{jk_{d+1}})$ means that the d-dimensional integrand function $V(x)$ is calculated for the set $y_{jk_1}, \ldots, y_{jk_d}$.

In Table B.1 we present the different ordered orbits $S_{[i]}, i = 0, \ldots, M_d$, for $d = 2, \ldots, 6$ that have been used in [1], where $M_2 = 2, M_3 = 4, M_4 = 6, M_5 = 10, M_6 = 14$. The orbit $S_{[i]} \equiv S_{m_1 \ldots m_{r_{di}}}$ contains the BC

$$(y_1, \ldots, y_{d+1}) = (\overbrace{\lambda_1, \ldots, \lambda_1}^{m_1 \text{ times}}, \ldots, \overbrace{\lambda_{m_{r_{di}}}, \ldots, \lambda_{m_{r_{di}}}}^{m_{r_{di}} \text{ times}}), \tag{B.4}$$

with

$$\sum_{j=1}^{r_{di}} m_j = d + 1, \quad \sum_{j=1}^{r_{di}} m_j \lambda_j = 1, \quad m_1 \geq \cdots \geq m_{r_{di}}. \tag{B.5}$$

The number of different permutations of the BC (B.4) is expressed by permutations of multisets

$$P_{di} = \frac{(d+1)!}{m_1! \cdots m_{r_{di}}!}. \tag{B.6}$$

The following formula holds for any permutations $(l_1, \ldots, l_{d+1})$ of $(k_1, \ldots, k_{d+1})$:

$$\int_{\Delta_d} x_1^{l_1} \cdots x_{d+1}^{l_{d+1}} dx = \frac{\prod_{i=1}^{d+1} k_i!}{\left(d + \sum_{i=1}^{d+1} k_i\right)!}, \quad x_{d+1} = 1 - \sum_{i=1}^{d} x_i. \tag{B.7}$$

Substituting symmetric polynomials with respect to the variables $x_1, \ldots, x_{d+1}$ of degree not exceeding p in (B.3) instead of $V(x)$, and taking into account (B.7), we obtain the system of nonlinear algebraic equations:

$$\int_{\Delta_d} s_2^{l_2} \times \cdots \times s_{d+1}^{l_{d+1}} dx = \frac{1}{d!} \sum_{i=0}^{M_d} P_{di} \sum_{j=1}^{K_{di}} W_{i,j} s_{i,j2}^{l_2} \times \cdots \times s_{i,jd+1}^{l_{d+1}}, \tag{B.8}$$

$$2l_2 + \cdots + (d+1)l_{d+1} \leq p, \tag{B.9}$$

where, for each orbit $S_{[i]}$, a set of K_{di} different BC is used. In (B.8), $K_{d0} = 0$ or 1 and $K_{di} \geq 0, i \neq 0$; $W_{i,j}$ is the jth weight of the orbit $S_{[i]}$;

$$s_k = \sum_{l=1}^{d+1} x_l^k, \quad k = 2, \ldots, d+1, \tag{B.10}$$

is the symmetric polynomial of degree k. Finally,

Table B.1 The orbits $S_{[i]} \equiv S_{m_1\cdots m_{r_{di}}}$ with different parameters r_{di} and their numbers of permutations P_{di} for $d = 2, \ldots, 6$

	$d=2$			$d=3$			$d=4$			$d=5$			$d=6$		
i	Orbits	r_{2i}	P_{2i}	Orbits	r_{3i}	P_{3i}	Orbits	r_{4i}	P_{4i}	Orbits	r_{5i}	P_{5i}	Orbits	r_{6i}	P_{6i}
0	S_3	1	1	S_4	1	1	S_5	1	1	S_6	1	1	S_7	1	1
1	S_{21}	2	3	S_{31}	2	4	S_{41}	2	5	S_{51}	2	6	S_{61}	2	7
2	S_{111}	3	6	S_{22}	2	6	S_{32}	2	10	S_{42}	2	15	S_{52}	2	21
3				S_{211}	3	12	S_{311}	3	20	S_{33}	2	20	S_{43}	2	35
4				S_{1111}	4	24	S_{221}	3	30	S_{411}	3	30	S_{511}	3	42
5							S_{2111}	4	60	S_{321}	3	60	S_{421}	3	105
6							S_{11111}	5	120	S_{222}	3	90	S_{331}	3	140
7										S_{3111}	4	120	S_{322}	3	210
8										S_{2211}	4	180	S_{4111}	4	210
9										S_{21111}	5	360	S_{3211}	4	420
10										S_{111111}	6	720	S_{2221}	4	630
11													S_{31111}	5	840
12													S_{22111}	5	1260
13													S_{211111}	6	2520
14													$S_{1111111}$	7	5040

$$s_{i,jk} = \sum_{l=1}^{r_{di}} m_l \lambda_{i,jl}^k \tag{B.11}$$

denotes the jth value of (B.10) on the components $\lambda_{i,jl}$ of the BC of the orbit $S_{[i]}$.

The count of all the solutions of Eq. (B.9) characterized by $l_k \geq 0$ provides the number E_{dp} of the independent nonlinear equations for the p-ordered quadrature rule. The value of E_{dp} can be calculated by recurrence for arbitrary d and p [1]:

$$E_{dp} = \begin{cases} 1 + \left\lfloor \frac{p}{2} \right\rfloor, & d = 1, \quad p \geq 0 \\ E_{d-1p}, & d \geq 2, \quad 0 \leq p \leq d, \\ E_{d-1p} + E_{dp-d-1}, & d \geq 2, \quad p \geq d+1, \end{cases} \tag{B.12}$$

where $\lfloor x \rfloor$ denotes the integer part of x.

Taking into account the conditions (B.8) at $l_2 = \cdots = l_{d+1} = 0$ and (B.5), the jth weight $W_{i,j}$ and components $\lambda_{i,jl}$ of the BC of the orbit $S_{[i]}$, must obey the simple bounds and the linear constraints respectively,

$$0 \leq W_{i,j} \leq \frac{1}{P_{di}}, \quad 0 \leq \lambda_{i,jl} \leq \frac{1}{m_l}, \quad 0 \leq \sum_{l=1}^{r_{di}-1} m_l \lambda_{i,jl} \leq 1. \tag{B.13}$$

The system of large number of nonlinear equations (B.8), (B.9) with the linear constraints (B.13) is solved by a modified Levenberg-Marquardt method [2–5]. Calculated minimal numbers N_{dp} of the nodes for PI-type fully symmetric p-order quadrature rules [6] are presented in Table B.2. For comparison, similar previously reported the numbers N_{dp} are also shown.

B.2   Estimates of the Errors of the Quadrature Rules

To estimate the error of the quadrature rules (B.1), we decompose the integrand $V(x)$ into a Taylor series in the vicinity of the point $x_t = (x_{1t}, \ldots, x_{dt})$ inside the simplex

$$V(x) = V^t(x) + O(x^{p+2}), \tag{B.14}$$

$$V^t(x) = \sum_{i_1 + \cdots + i_d \leq p+1} V^{(i_1, \ldots, i_d)}(x_t) \frac{(x_1 - x_{1t})^{i_1} \times \cdots \times (x_d - x_{dt})^{i_d}}{i_1! \times \cdots \times i_d!},$$

where $V^{(i_1, \ldots, i_d)}(x_t)$ is a mixed derivative at $x = x_t$ and consider the auxiliary function

$$\varepsilon(V(x)) = \left| \int_{\Delta_d} V(x)dx - \frac{1}{d!} \sum_{j=1}^{N_{dp}} w_j V(x_{j1}, \ldots, x_{jd}) \right|. \tag{B.15}$$

Table B.2 The minimal numbers N_{dp} of nodes for PI-type fully symmetric p-order quadrature rules and comparison with the known numbers N_{dp}

N_{dp}

p	$d = 2$		$d = 3$					$d = 4$			$d = 5$		$d = 6$	
	[6-8]	[9]	[6]	[10]	[8]	[9]	[11]	[6]	[10]	[12]	[6]	[10]	[6]	[10]
4	6	6	14	14	14	14	14	20	20	20	27	27	43	43
5	7	7	14	14	14	14	14	30	30	30	37	37	64	64
6	12	12	24	24	24	24	24	56	56	56	102	102	175	175
7	15	15	35	35	35	36	35	70	70	70	137	137	252	266
8	16	16	46	46	46	46	46	105	105	105	228	257	448	553
9	19	19	59		59	61	59	151	110	151	338		700	
10	25	25	79		81	81	81	210		210	479		1078	
11	28	28	98			109	110	275		281				
12	33	33	123			140	168	370		445				
13	37	37	145			171	172	470		555				
14	42	46	175			236	204	601		725				
15	49	52	209				264	781		905				
16	55	55	248				304	956		1055				
17	60	61	284				364							
18	67	72	343				436							
19	73	73	383				487							
20	79	88	441				552							

Table B.3 The list of quadrature rules on triangle with the corresponding combination of orbits and their error estimates

p	N_{dp}	S_3	S_{21}	S_{111}	$\max \varepsilon_{i_1,i_2}$	$\sum \varepsilon_{i_1,i_2}$	$\sqrt{\sum \varepsilon_{i_1,i_2}^2}$
4	6		2		$2.62 \cdot 10^{-6}$	$4.71 \cdot 10^{-6}$	$3.00 \cdot 10^{-6}$
5	7	1	2		$1.27 \cdot 10^{-6}$	$2.64 \cdot 10^{-6}$	$1.61 \cdot 10^{-6}$
6	12		2	1	$2.46 \cdot 10^{-9}$	$6.28 \cdot 10^{-9}$	$3.44 \cdot 10^{-9}$
7	15		1	2	$3.82 \cdot 10^{-9}$	$8.96 \cdot 10^{-9}$	$5.15 \cdot 10^{-9}$
8	16	1	3	1	$3.89 \cdot 10^{-11}$	$9.01 \cdot 10^{-11}$	$5.00 \cdot 10^{-11}$
9	19	1	4	1	$6.30 \cdot 10^{-12}$	$2.10 \cdot 10^{-11}$	$1.16 \cdot 10^{-11}$
10	25	1	2	3	$9.39 \cdot 10^{-15}$	$3.00 \cdot 10^{-14}$	$1.48 \cdot 10^{-14}$
11	28	1	5	2	$3.08 \cdot 10^{-15}$	$6.69 \cdot 10^{-15}$	$3.99 \cdot 10^{-15}$
12	33		5	3	$1.02 \cdot 10^{-17}$	$3.57 \cdot 10^{-17}$	$1.66 \cdot 10^{-17}$
13	37	1	4	4	$1.28 \cdot 10^{-17}$	$3.56 \cdot 10^{-17}$	$1.87 \cdot 10^{-17}$
14	42		6	4	$7.09 \cdot 10^{-20}$	$2.29 \cdot 10^{-19}$	$1.12 \cdot 10^{-19}$
15	49	1	4	6	$2.95 \cdot 10^{-21}$	$7.51 \cdot 10^{-21}$	$4.11 \cdot 10^{-21}$
16	55	1	4	7	$5.31 \cdot 10^{-23}$	$1.92 \cdot 10^{-22}$	$8.71 \cdot 10^{-23}$
17	60		6	7	$8.27 \cdot 10^{-24}$	$2.66 \cdot 10^{-23}$	$1.29 \cdot 10^{-23}$
18	67	1	6	8	$1.24 \cdot 10^{-26}$	$3.92 \cdot 10^{-26}$	$1.94 \cdot 10^{-26}$
19	73	1	6	9	$3.73 \cdot 10^{-28}$	$1.13 \cdot 10^{-27}$	$5.14 \cdot 10^{-28}$
20	79	1	8	9	$5.92 \cdot 10^{-30}$	$2.09 \cdot 10^{-29}$	$9.65 \cdot 10^{-30}$

Taking into account that the quadrature rule is exact for polynomials of degree less than p, one has

$$
\begin{aligned}
\varepsilon(V^t(x)) &= \left| \int_{\Delta_d} V^t(x)dx - \frac{1}{d!} \sum_{j=1}^{N_{dp}} w_j V^t(x_{j1}, \ldots, x_{jd}) \right| \\
&= \left| \sum_{i_1+\cdots+i_d=p+1} V^{(i_1,\ldots,i_d)}(x_t) \left(\int_{\Delta_d} \frac{x_1^{i_1} \times \cdots \times x_d^{i_d}}{i_1! \times \cdots \times i_d!} dx - \frac{1}{d!} \sum_{j=1}^{N_{dp}} w_j \frac{x_{j1}^{i_1} \times \cdots \times x_{jd}^{i_d}}{i_1! \times \cdots \times i_d!} \right) \right| \\
&\leq \sum_{i_1+\cdots+i_d=p+1} |V^{(i_1,\ldots,i_d)}| \varepsilon_{i_1,\ldots,i_d}, \quad \varepsilon_{i_1,\ldots,i_d} \equiv \varepsilon\left(\frac{x_1^{i_1} \times \cdots \times x_d^{i_d}}{i_1! \times \cdots \times i_d!} \right),
\end{aligned}
\tag{B.16}
$$

where $|V^{(i_1,\ldots,i_d)}|$ is the absolute maximum value of the mixed derivative on the simplex. As it can be seen from (B.16), to estimate the errors of quadrature rules, it is enough to calculate the coefficients $\varepsilon_{i_1,\ldots,i_d}$ for the corresponding derivatives. However, there are quite

Table B.4 The same as in Table B.3, but for the tetrahedron

p	N_{dp}	S_4	S_{31}	S_{22}	S_{211}	S_{1111}	$\max \varepsilon_{i_1,i_2,i_3}$	$\sum \varepsilon_{i_1,i_2,i_3}$	$\sqrt{\sum \varepsilon_{i_1,i_2,i_3}^2}$
4	14		2	1			$4.73 \cdot 10^{-8}$	$1.42 \cdot 10^{-7}$	$6.95 \cdot 10^{-8}$
5	14		2	1			$1.43 \cdot 10^{-7}$	$3.98 \cdot 10^{-7}$	$1.96 \cdot 10^{-7}$
6	24		3		1		$4.82 \cdot 10^{-9}$	$1.60 \cdot 10^{-8}$	$6.92 \cdot 10^{-9}$
7	35	1	1	1	2		$1.86 \cdot 10^{-10}$	$9.95 \cdot 10^{-10}$	$3.71 \cdot 10^{-10}$
8	46		4	1	2		$2.80 \cdot 10^{-11}$	$6.09 \cdot 10^{-11}$	$3.24 \cdot 10^{-11}$
9	59	1	4	1	3		$6.19 \cdot 10^{-13}$	$2.03 \cdot 10^{-12}$	$8.83 \cdot 10^{-13}$
10	79	1	3	1	5		$1.47 \cdot 10^{-14}$	$8.40 \cdot 10^{-14}$	$2.68 \cdot 10^{-14}$
11	98		5	1	4	1	$8.59 \cdot 10^{-16}$	$3.39 \cdot 10^{-15}$	$1.21 \cdot 10^{-15}$
12	123	1	5	1	6	1	$8.73 \cdot 10^{-18}$	$6.79 \cdot 10^{-17}$	$2.00 \cdot 10^{-17}$
13	145	1	3	2	8	1	$8.01 \cdot 10^{-19}$	$3.27 \cdot 10^{-18}$	$1.21 \cdot 10^{-18}$
14	175	1	6	1	10	1	$2.23 \cdot 10^{-20}$	$1.11 \cdot 10^{-19}$	$3.49 \cdot 10^{-20}$
15	209	1	4	2	11	2	$5.92 \cdot 10^{-22}$	$4.32 \cdot 10^{-21}$	$1.14 \cdot 10^{-21}$
16	248		8	2	11	3	$1.04 \cdot 10^{-23}$	$9.06 \cdot 10^{-23}$	$2.25 \cdot 10^{-23}$
17	284		8	2	14	3	$5.29 \cdot 10^{-25}$	$3.30 \cdot 10^{-24}$	$8.96 \cdot 10^{-25}$
18	343	1	6	1	18	4	$2.02 \cdot 10^{-27}$	$1.93 \cdot 10^{-26}$	$4.58 \cdot 10^{-27}$
19	383	1	7	3	18	5	$1.37 \cdot 10^{-28}$	$1.50 \cdot 10^{-27}$	$3.39 \cdot 10^{-28}$
20	441	1	8	4	20	6	$8.94 \cdot 10^{-30}$	$4.32 \cdot 10^{-29}$	$1.32 \cdot 10^{-29}$

a lot of such coefficients, so to compare the quadrature rules found, we limited ourselves to the largest of the coefficients, $\max \varepsilon_{i_1,\dots,i_d}$, their sum $\sum \varepsilon_{i_1,\dots,i_d}$ and the root of the sum of their squares $\sqrt{\sum \varepsilon_{i_1,\dots,i_d}^2}$, where summation was carried out over sets of numbers $i_1, \dots, i_d$ at $i_1 + \cdots + i_d = p + 1$ and only one permutation $i_1, \dots, i_d$ was taken into account. The results obtained are presented in the Tables B.3, B.4, B.5, B.6 and B.7.

Table B.5 The same as in Table B.3, but for the 4-simplex

p	N_{dp}	S_5	S_{41}	S_{32}	S_{311}	S_{221}	S_{2111}	$\max \varepsilon_{i_1,\dots,i_4}$	$\sum \varepsilon_{i_1,\dots,i_4}$	$\sqrt{\sum \varepsilon_{i_1,\dots,i_4}^2}$
4	20		2	1				$6.23 \cdot 10^{-8}$	$2.88 \cdot 10^{-7}$	$1.20 \cdot 10^{-7}$
5	30		2	2				$1.97 \cdot 10^{-8}$	$6.70 \cdot 10^{-8}$	$2.76 \cdot 10^{-8}$
6	56	1	1	1	2			$2.63 \cdot 10^{-10}$	$1.23 \cdot 10^{-9}$	$4.28 \cdot 10^{-10}$
7	70		2	2	2			$1.03 \cdot 10^{-10}$	$1.95 \cdot 10^{-10}$	$1.08 \cdot 10^{-10}$
8	105		3	2	2	1		$2.15 \cdot 10^{-12}$	$5.70 \cdot 10^{-12}$	$2.49 \cdot 10^{-12}$
9	151	1	2	2	3	2		$4.03 \cdot 10^{-14}$	$2.26 \cdot 10^{-13}$	$6.73 \cdot 10^{-14}$
10	210		4	2	4	3		$1.59 \cdot 10^{-15}$	$9.73 \cdot 10^{-15}$	$2.79 \cdot 10^{-15}$
11	275		3	3	4	3	1	$4.12 \cdot 10^{-17}$	$3.61 \cdot 10^{-16}$	$8.71 \cdot 10^{-17}$
12	370		4	4	5	3	2	$1.06 \cdot 10^{-18}$	$9.59 \cdot 10^{-18}$	$1.97 \cdot 10^{-18}$
13	470		4	5	5	4	3	$4.66 \cdot 10^{-20}$	$3.91 \cdot 10^{-19}$	$9.38 \cdot 10^{-20}$
14	601	1	4	2	10	6	3	$1.42 \cdot 10^{-21}$	$1.36 \cdot 10^{-20}$	$2.89 \cdot 10^{-21}$
15	781	1	4	2	10	8	5	$7.88 \cdot 10^{-23}$	$3.35 \cdot 10^{-22}$	$9.21 \cdot 10^{-23}$
16	956	1	5	4	10	9	7	$2.79 \cdot 10^{-25}$	$4.29 \cdot 10^{-24}$	$7.67 \cdot 10^{-25}$

Table B.6 The same as in Table B.3, but for the 5-simplex

p	N_{dp}	S_6	S_{51}	S_{42}	S_{33}	S_{411}	S_{321}	S_{222}	$\max \varepsilon_{i_1,\dots,i_5}$	$\sum \varepsilon_{i_1,\dots,i_5}$	$\sqrt{\sum \varepsilon_{i_1,\dots,i_5}^2}$
4	27	1	1		1				$6.89 \cdot 10^{-8}$	$1.11 \cdot 10^{-7}$	$7.24 \cdot 10^{-8}$
5	37	1	1	2					$1.39 \cdot 10^{-9}$	$4.79 \cdot 10^{-9}$	$2.10 \cdot 10^{-9}$
6	102		2	2			1		$6.31 \cdot 10^{-11}$	$2.22 \cdot 10^{-10}$	$8.12 \cdot 10^{-11}$
7	137		2	1	1	1	1		$1.23 \cdot 10^{-12}$	$9.20 \cdot 10^{-12}$	$2.76 \cdot 10^{-12}$
8	228	1	2	1	1	2	2		$1.34 \cdot 10^{-13}$	$4.92 \cdot 10^{-13}$	$1.68 \cdot 10^{-13}$
9	338		3	2	1	1	4		$1.42 \cdot 10^{-14}$	$2.72 \cdot 10^{-14}$	$1.50 \cdot 10^{-14}$
10	479		4	1	1	3	4	1	$8.36 \cdot 10^{-17}$	$6.05 \cdot 10^{-16}$	$1.51 \cdot 10^{-16}$

Table B.7 The same as in Table B.3, but for the 6-simplex

p	N_{dp}	S_7	S_{61}	S_{52}	S_{43}	S_{511}	S_{421}	S_{331}	S_{322}	$\max \varepsilon_{i_1,\dots,i_6}$	$\sum \varepsilon_{i_1,\dots,i_6}$	$\sqrt{\sum \varepsilon_{i_1,\dots,i_6}^2}$
4	43	1	1		1					$2.23 \cdot 10^{-9}$	$4.88 \cdot 10^{-9}$	$2.54 \cdot 10^{-9}$
5	64	1	1	1	1					$2.45 \cdot 10^{-10}$	$5.89 \cdot 10^{-10}$	$2.89 \cdot 10^{-10}$
6	175	2	1	1			1			$2.41 \cdot 10^{-12}$	$1.43 \cdot 10^{-11}$	$4.45 \cdot 10^{-12}$
7	252	2	1	2	1	1				$3.01 \cdot 10^{-13}$	$8.26 \cdot 10^{-13}$	$3.38 \cdot 10^{-13}$
8	448	3		2	1	3				$7.08 \cdot 10^{-15}$	$3.30 \cdot 10^{-14}$	$1.04 \cdot 10^{-14}$
9	700	2	2	3	2	1	1		1	$6.48 \cdot 10^{-16}$	$2.05 \cdot 10^{-15}$	$7.34 \cdot 10^{-16}$
10	1078	2	3	2	3	3	2	1		$3.99 \cdot 10^{-18}$	$3.30 \cdot 10^{-17}$	$7.19 \cdot 10^{-18}$

References

1. J.I. Maeztu, E. Sainz de la Maza, Consistent structures of invariant quadrature rules for the n-simplex, Math. Comput. **64**, 1171–1192 (1995)
2. K. Levenberg, A method for the solution of certain non-linear problems in least squares, Q. Appl. Math. **2**, 164–168 (1944)
3. D. Marquardt, An algorithm for least squares estimation of parameters, J. Soc. Ind. Appl. Math. **11**, 431–441 (1963)
4. Ch. Kanzow, N. Yamashita, M. Fukushima, Levenberg–Marquardt methods with strong local convergence properties for solving nonlinear equations with convex constraints, J. Comput. Appl. Math. **172**, 375–397 (2004)
5. N. Marumo, T. Okuno, A. Takeda, Majorization-minimization-based Levenberg–Marquardt method for constrained nonlinear least squares, arXiv:2004.08259v2, pp. 1–32 (2021)
6. G. Chuluunbaatar, O. Chuluunbaatar, A.A. Gusev, S.I. Vinitsky, PI-type fully symmetric quadrature rules on the 3-, …, 6-simplexes, Comput. Math. Appl. **124**, 89–97 (2022).
7. H. Xiao, Z. Gimbutas, A numerical algorithm for the construction of efficient quadrature rules in two and higher dimensions, Comput. Math. Appl. **59**, 663–676 (2010)
8. F.D. Witherden, P.E. Vincent, On the identification of symmetric quadrature rules for finite element methods, Comput. Math. Appl. **69**, 1232–1241 (2015)
9. L. Zhang, T. Cui, H. Liu, A set of symmetric quadrature rules on triangles and tetrahedra, J. Comput. Math. **27**, 89–96 (2009)
10. A.A. Gusev, V.P. Gerdt, O. Chuluunbaatar, G. Chuluunbaatar, S.I. Vinitsky, V.L. Derbov, A. Góźdź, P.M. Krassovitskiy, Symbolic-numerical algorithms for solving elliptic boundary-value problems using multivariate simplex Lagrange elements, Lect. Notes Comput. Sci. **11077**, 197–213 (2018)

11. J. Jaśkowiec, N. Sukumar, High-order symmetric cubature rules for tetrahedra and pyramids, Int. J. Numer. Methods Eng. **122**, 148–171 (2021)
12. C.V. Frontin, G.S. Walters, F.D. Witherden, W. Lee, D.M. Williams, D.L. Darmofal, Foundations of space-time finite element methods: polytopes, interpolation, and integration, Appl. Numer. Math. **166**, 92–113 (2021)